Secrets of Recording

Secrets of Recording: Professional Tips, Tools & Techniques

Lorne Bregitzer

Routledge
Taylor & Francis Group
New York London

First published 2009

This edition published 2013
by Focal Press
70 Blanchard Road, Suite 402, Burlington, MA 01803

Simultaneously published in the UK
by Focal Press
2 Park Square, Milton Park, Abingdon, Oxon OX14 4RN

First issued in hardback 2017

Focal Press is an imprint of the Taylor & Francis Group, an informa business

Library of Congress Cataloging-in-Publication Data
Bregitzer, Lorne.
  Secrets of recording : professional tips, tools & techniques / Lorne Bregitzer.
    p. cm.
  Includes index.
  ISBN 978-0-240-81127-7 (pbk. : alk. paper) 1. Sound—Recording and reproducing—Digital
techniques. 2. Sound studios. I. Title.
  TK7881.4.B74 2007
  781.49–dc22   2008041808

British Library Cataloguing-in-Publication Data
A catalogue record for this book is available from the British Library.

ISBN 13: 978-0-2408-1127-7 (pbk)
ISBN 13: 978-1-1384-6898-6 (hbk)

Contents

CHAPTER 1 • Creating a Great Recording from the Start

CHAPTER 2 • Timing Correction

CHAPTER 3 • Replacing Sounds

CHAPTER 4 • Pitch Correction

CHAPTER 5 • Emulated Effects

CHAPTER 6 • Adding MIDI Tracks to Recordings

CHAPTER 7 • Mixing Techniques

CHAPTER 8 • Mastering the Recording

Acknowledgments

I would first like to thank Sam McGuire for originally coming up with the idea for this book.

Thanks to my perfect wife, Marla, for all the patience and support that she has given me throughout the writing of this book, as well as throughout my career.

Many thanks to my Mom and Dad, Max & Lee Bregitzer, as well as my sister, Karen, for supporting me all throughout my chosen career in addition to the support that they have given me while writing this book.

Thanks to those who have helped me in my career. First, Kevin Clock at Colorado Sound Studios for giving me my first break in this industry. Thanks to everyone else at Colorado Sound for their knowledge and help throughout the years, including Tom Capek, JP Manza, Jessie O'Brien, Steve Avedis, Tammy Baretta, Iam Hlatky, and Cheryl Winston.

Thanks to everyone at the University of Colorado Denver. Thanks to Rich Sanders for giving me my first break at the university. Thanks to the recording arts faculty: Leslie Gaston, Tom Lang, Sam McGuire, and the late Roy Pritts. Thanks to David Dynak for helping me along the way to the next step of my career as well as Frank Jermance for your guidance in helping me to get there.

Thanks to everybody at Focal Press, including Catharine Steers and Carlin Reagan, as well as the countless others who have worked on this book.

I would like to thank Joy Caylor for helping to foster an appreciation of music and having a tremendous impact on my life.

I would also like to thank all of the artists that I have had the pleasure to work with.

Thanks to all of my students for their energy and interest.

Thanks to all of the producers and engineers that I have had the pleasure to observe and learn from.

I would especially like to thank all of the teachers and instructors that I have had over the years. Every one of them has left a positive impact on my life.

The intention of this book is to move beyond the wealth of audio textbooks and provide you with contemporary audio production techniques where the other books leave off. If you are looking to become a technologically advanced engineer, this book will supplement a solid understanding of audio engineering provided by many of the other outstanding textbooks on the subject.

Throughout recorded history, there has been much experimentation with what can be manipulated in the world of audio. Musicians, engineers, and music fans, in general, have all uttered the phrase, "How'd they do that?"

This book will give you an overview of many of the tools and tricks used today. Many modern records are much more "perfect" today than they were in previous decades. With this perfection of tempo, time, and technique, many up-and-coming engineers and producers rely on these tools to make their records competitive with what is being played on the radio.

Many prospective engineers will need to know these tools, as their clients will often request them. They may not know the exact terminology, but they will say, "Can't you fix that?"

I cannot count the number of times I have had a client ask me in the studio, "Can you do that Cher effect on my vocal?" Like it, or not, you had better be able to create the effect for them so you can talk them out of it, if possible.

Many more effects are transparent to the client, such as pitch and timing correction. Being able to create these effects in the studio will give your musicians the competitive edge that they are looking for, and this will make for an impressed client and an outstanding final product.

Many of these techniques are tools for you to learn. It is like being a mechanic or a carpenter. Each of these people has a variety of tools at his or her disposal and knows when to use the right tool at the right time. You would not hire a contractor to remodel your kitchen who is only planning on walking into your house with a screwdriver. The same holds true for the recording engineer; the more tools that you have at your disposal, the more versatile an engineer you will be.

Knowing these tools is important, and you have to become fluent with them. You cannot just buy the software and then suddenly have these abilities. Would you hire a musician for a session who claims to be a guitar player simply because he or she bought a guitar last week? Of course not. You need to practice with these tools so they will become a seamless part of your repertoire. Would a client be confident in your abilities if you had to break out the manual when you are setting up microphones so that you know what the polar pattern of the microphone is, or if it requires phantom power? The answer is, obviously, no. Similarly, a client is not going to be happy with you during your recording session when you have to break out the manual to figure out how to correct the pitch of a vocal track.

Certain aspects of this book may not be for everybody. If you are looking to record and capture sounds entirely naturally, then some of the technological tricks may not be something of interest to you. These techniques can be used with all styles of music, however. Even if you are just focusing on acoustic folk music and do not think that you will ever use any of these tricks, you will be surprised by what you can do by merely being able to tighten up background vocals to the lead vocal, or to correct a few measures of a drum part.

Is it considered cheating? Many people consider the use of these techniques to be cheating by the musicians, engineers, and producers. These are really the "grumpy old men" of audio chasing the kids off their lawn with an Ampex tape machine instead of a broom. Many people in the industry do not consider this to be cheating. It should be noted that most engineers and musicians who employ these techniques do not go on record saying that they have used them. As a result, the public has the perception that everything was recorded entirely naturally.

The audience for music will not pay attention to the fact that every kick and snare drum sound are consistent—a little too consistent. They will not pay attention to the fact that the ten-second note held by the vocalist is precisely in pitch. They will notice if the vocals are drastically out of tune, or if the kick drum is inconsistent on every single hit.

Manufactured recording has been around for decades. How often have you read about bands spending over a year in the studio? This is not because they were recording 200 tracks of guitar to be used on a single song. It is because they have been employing the techniques of their day to come up with the perfect recording. Modern techniques are employing the same principles, but going about them differently.

Today's techniques are nowhere near as time consuming and expensive as they were in previous decades. With the power of today's computers coupled with the relatively inexpensive nature of some of these tools, a $1200 laptop has the capability of doing what these previous artists spent hundreds of thousands of dollars doing.

By using these techniques, you are not necessarily cheating. If employed correctly, you can improve a vocal performance that had the energy and emotion that you were looking to capture, but it may have had some issues with intonation. Correcting the pitch, rather than using a lesser-quality performance, will give you the best possible performance from your vocalist.

You certainly can take musicians and make them sound better than they are. Is the utilization of modern correction techniques any different than in the past? As an engineer, you used to have to make the musicians play their part dozens of times until they got it right. You might have captured the 1 take out of 100 that they performed correctly, or recorded 3 takes, and then corrected the timing to make it work in the record. The musician's inadequacy is not reflected, in either case, in the finished recording. Most consumers want a record with good songs and good performances. Most bands are looking for the same thing.

Sometimes musicians are hesitant to use any tricks for their recording, but oftentimes, if you can demonstrate to them how it will enhance their record, they will usually be excited about these new techniques. As an engineer you will oftentimes have to educate your clients about the recording process, and what you, as an engineer, can do for them in the studio.

This book will go beyond the techniques of engineering taught by most books. It will impart the reader with tips and tricks for recording that are used by engineers who have gained these techniques from years of experience of trial and error. As an engineer and an educator, I try to go beyond what is taught in textbooks and in the classroom and impart the techniques, tools, and tricks that I have used successfully with clients for more than a decade.

No single engineer knows all of the tricks and techniques used in the world of audio production. Being able to observe and learn from other engineers around you helps you to build up a solid repertoire of engineering skills, techniques, and tricks.

Being a successful engineer goes beyond understanding the principles of acoustics, microphone technique, and digital audio. To be successful, you need to be able to make your clients comfortable, develop an ear for audio, and know how to work a studio session seamlessly. You need to keep current with technology to compete with other engineers. Bands are oftentimes looking for the best deal they can get for their budget. If you can convince them that you have the knowledge and tools to make the most of their recording, you will stand above somebody who has spent a few hundred dollars at their local music store looking to record a band for a bargain.

An up-to-date studio engineer will demonstrate value to his or her clients, as he or she can work with the clients' sounds and make the most out of their studio time. If you charge twice as much as a cheaper and less knowledgeable engineer, you can demonstrate to them that they will get twice as much accomplished with you per hour than if they went elsewhere. Additionally, they will be more satisfied with the end result. The key to doing this is becoming well versed in the techniques of today with an eye for what is coming up tomorrow.

With this book, you will learn many of the techniques in dealing with musicians. But more important, you will learn how to make the most of your studio time by learning numerous tricks that successful musicians, engineers, and producers use today.

CHAPTER 1

Creating a Great
Recording from the Start

FINE-TUNING YOUR STUDIO

When you are putting together and building your home or project studio, there are many things you can do to improve the quality of recorded sound. Having the best equipment does not necessarily capture the best quality of audio. Knowing how to make the most out of your acoustic space and minimizing any problems will benefit the tracks in the long run.

Choosing appropriate monitors

When first putting a studio together, many first-time engineers will make the mistake of not choosing suitable monitors. There are places in the studio where you can save money, but the proper choice of monitors is going to affect the quality of sound from the tracking through the mixing. These are speakers that you're going to have to listen to, for hours at a time. Monitors are classified as near-field, mid-field, or far-field. This designates the distance of the listener from the monitors. Most home studios will utilize near-field monitors, as you may not have enough space in the control room to place mid-field or far-field monitors.

POWERED VERSUS UNPOWERED MONITORS

There are two types of monitors that you can select for your studio. The first is the standard unpowered monitor. This will require a separate amplifier to bring the line level up to speaker level. Unpowered monitors allow you to choose which sort of application you wish to use with them. They require more cabling, but many monitors used in professional studios are unpowered.

More recently, power monitors have become standard in many studios. These will have the application built in, so they will only require a line-level signal. Many of these powered monitors will even accept a digital input, so it will take care of the analog to digital conversion as well as amplification. These monitors can accept both balanced and unbalanced signals, but they will also require A/C power to feed the amplifier built inside. Powered monitors will also have the ability to adjust the high and low frequencies by manipulating a few switches on the back. Some powered monitors are actually self-calibrating, such as the JBL LSR 4300 series, which will contour the sound based on the way the speakers sound in the room. These models will often come with a dedicated microphone and software with which to perform the calibration.

Calibration for each of these different types of monitors can be done with an external equalizer. A graphic equalizer can be used to fine-tune the sound of a room; however, you can do more damage to the sound of the room if you do not know what you are doing. If you are in doubt, leave the fine-tuning of the monitors alone, and make adjustments only to the high-frequency and low-frequency adjustments of a powered monitor.

PLACEMENT OF THE MONITORS

The placement of monitors, in a studio, can make a dramatic difference in sound to the engineer. Monitors should be placed carefully, such that the two monitors and the engineer form an equilateral triangle. This means that there should be a 60-degree angle between the two monitors as well as from each monitor to the next. If you are looking straight ahead, the monitor should be 30 degrees to the right and to the left of the center.

Monitors will have multiple drivers in them. There can be midrange drivers as well as tweeters. Since high frequencies are more directional, be sure that these drivers are directed toward the listening position at ear level. Depending on the layout of the studio, there can be issues with reflecting high frequencies off flat surfaces. These can come off a desk or control surface in front of the engineer. It is best to angle the speakers in such a way that they do not create too many direct reflections off the surfaces.

CALIBRATING THE MONITORS

The monitors should sound balanced across low and high frequencies. Mixing to monitors that are lacking in high frequencies, whether it is a result of the acoustics or placement of the monitors, will result in mixes that are too bright. If the high frequencies are too hyped, then the mixes will come across as sounding dull. The same holds true for the low frequencies. If the low frequencies from the monitors are too loud, the mixes will be lacking in low-end, while the reverse is true for not having enough low-frequency gain. This is where the calibration of the monitors is important.

The calibration can be done by using a sound pressure level (SPL) meter, and by playing back a consistent level from your digital audio workstation (DAW). This can be done by creating an oscillator, with a sine wave at 1 kHz, which is the standard reference level, and taking a reading of that level. Then switch the oscillator to 10 kHz. If the reading is the same as the 1-kHz level, then the monitors do not need any adjustment on the high frequencies. If the readings do not match, then adjust the high-frequency playback on the monitors or a parametric equalizer that is placed across the monitor path. Do the same thing with 100 Hz.

More often than not, the calibration can be done by ear. When listening back to the mixes on a different sound system, if they are sounding a little too dull, then reduce the high frequencies of the monitors. Do the same thing with the low frequencies, and then listen to how your mixes come across after the adjustments.

Acoustic treatment of the control room

The control room is your first and primary source for listening to the quality of the recorded audio. Creating a neutral listening environment will help make the mixes sound as good as possible. Control rooms, in a home studio, are generally in rooms that are not ideal for listening back to professional audio, compared to a professionally designed space. Oftentimes they are in a basement, which may have hard tile floors, painted concrete walls, and low ceilings.

In a small space, the listening environment can be improved by absorbing much of the acoustic sound. The room does not necessarily need to sound completely dead, as no one will be listening to the music in a completely dead space. Concrete and tile are very reflective surfaces. Placing carpet, even an area rug, will absorb a good deal of the sound. The walls can be treated with professional acoustic foam. This can be expensive, but this investment will make it worthwhile in the long run. Wood is a semireflective surface. Placing wood along the walls is a good way of absorbing some of the sound, while maintaining a realistic listening environment. Bass traps can be placed in the corners to help eliminate or reduce any low-frequency room nodes that may exist in the control room.

ACOUSTICS IN THE STUDIO

Controlling the acoustics in the studio, while the recording is taking place, is important so that there is no unnecessary room noise that can detract from the sound of the recording. Professional recording studios will have different rooms. Some rooms are made out of wood, so that the room sound is semireflective. A room like this will sound good when recording most instruments. Some studios will also have a tiled room, which will contribute more reflections of the room into microphones. Recording drums in a tiled room will capture more reflections of the room. It is especially beneficial when recording the sound of the room itself. The third type of space used in a recording studio is an isolation booth (Figure 1.1). This is a dead-sounding room, designed to absorb most of the frequencies. These rooms are used when recording the voice for vocals or voiceovers. It can also be used for capturing soul instruments such as a saxophone, acoustic guitar, or any instrument that does not necessarily need to have the sound of the room to be picked up by the microphones.

FIGURE 1.1
An isolation room for single-instrument or vocal recording.

FIGURE 1.2
Portable gobos for isolating specific instruments in a room.

A home recording studio may not have the luxury of having separate rooms to record in. The acoustics of the room can be tailored with portable gobos that can be maneuvered around the room (Figure 1.2). Having a vocalist surrounded by these gobos, which are designed to absorb sound, will help prevent unwanted room sound from creeping into the microphone. The louder the vocalist sings, the more apt there is to be captured room noise in the microphone.

Heating and air conditioning can also create unwanted noise in both the studio and control room. It may not be feasible to reroute the heating and air conditioning in a home or project studio, so make sure that the heating and air conditioning is shut off when recording, especially with softer instruments, which will require more gain from the microphone preamplifiers.

WORKING WITH MUSICIANS

When making a record, one of the first and foremost things that you can do to have a great-sounding final product is to have the best musicians possible. Studio musicians are highly paid musicians because they can walk in and out of the studio, get their parts done, and make the producer happy with the end result. Most bands that come into the studio are not studio musicians, so it takes work to make them comfortable with the studio. The key to a successful recording is to prepare the musicians in advance.

Good preparation will help make a band make the most of their time while they are in the studio. As an engineer, you may not have the time or availability to work with a band as they rehearse for the studio, but what you can do is let them know what to expect when they walk in the door. If they have never been in a studio, meet with them and explain the process to them in advance. This may help to answer questions and relieve anxieties that an artist can feel before walking into the studio for a new project. There is no such thing as overpreparing for the recording studio. There are many ways in which you can prepare a band as a whole, as well as the individual musicians, for the recording process. This is especially true for bands that have never been in a studio before. Educating them in advance about the studio process is an important part of the process of making them feel comfortable with recording.

Preparing for a recording during rehearsal

It is very useful to work with an artist, if possible, during rehearsal and prior to recording, to make the most of your studio time.

THE ARRANGEMENT

Finalizing the arrangement of the song prior to entering the studio is very important. If you take an artist into the studio and then start rearranging the songs, the artist is not making the most out of his or her studio time.

For many reasons, it is helpful to do a simple recording of the band as a whole during rehearsal. A simple two-track recording using something like a CD recorder or MP3 recorder can suffice. Many musicians are more focused on their parts during rehearsal and not listening to the performance as a whole. This gives the band the opportunity to point out trouble spots, whether they are problems with the arrangement or tempo fluctuations during various parts of the song.

Recording the song will also tell you the length of it and whether the duration fits within the band's appropriate goals for that song. If you are looking to have a radio-friendly pop song, then a six-minute song with a one-minute guitar solo may need some rearranging.

CHOOSING TO USE A CLICK TRACK AND TEMPO

If you are planning on recording to a click track, this is the ideal time to find the tempo of each of the songs that you will be recording.

Since most bands do not rehearse with a click track, you may find that their songs may fluctuate in tempo. Sometimes the choruses may jump up a little in speed, as opposed to the verses. If you are looking to have a solid single tempo for the songs, you will have to find a compromise. Sometimes it feels best to match the tempo to what sounds best in the choruses, and sometimes a compromise between the two tempos would be best. Not deciding on a tempo in advance can lead to bands regretting the tempo that they chose in the studio, as they may be too busy and concerned with other things.

CHORD CHART

It is helpful to put together a chord chart for each song. This will allow you to easily prepare for any pitch correction that may be necessary. If you do not have the time to put these charts together, at least take note of the key for each of the songs, which will help in the future. There is a handy trick for determining the chords of a song that has already been recorded, which will be discussed in Chapter 4 in the "Tracking the pitch of a bass track using auto-tune" section.

COMING UP WITH A PLOT FOR THE RECORDING SESSION

One of the differences between rehearsing and recording that you may want to prepare your clients for is the arrangement of the space. This becomes crucial in capturing the sound as effectively and efficiently as possible. If the band members are going to be in separate rooms, help them prepare and let them know that they may not have the same eye contact as they would if they were rehearsing.

Working with musicians in a rehearsal setting or meeting with them beforehand can help you come up with a game plan for the recording. If the band you are working with describes their music as rock, go in deeper and find what style of rock music they play, and what sound they are looking to come out of the studio with. Are they looking for a raw sound like Nirvana, or do they want a more produced rock sound, along the lines of Evanescense? This will help you decide how to choose which microphones to use, as well as how to best position the performers in the studio.

If the band is after a particular sound, ask them to give some examples. What recordings out there appropriately capture the sound that they are looking for? If the band gives you some recording examples, do some research as to what the engineer on that project used. There are many articles and interviews online where engineers are more than happy to share at least some of the techniques that they used.

What the musicians should bring into the studio

There are always several things the musicians should bring when coming to the studio. The more experienced studio musicians are already aware of what they need to bring. The inexperienced musicians will appreciate any advice that will help them save time and make the studio a more enjoyable experience.

Items musicians should bring:

- Extra strings (full sets, bass and guitar).
- Drum sticks.
- Drum key.
- Extra drum heads.
- Different snare drums (if available).
- Tuners for stringed instruments.
- Snacks.
- Beverages (nonalcoholic, obviously).
- Lyric sheets.
- Song charts (helpful, but not necessary).
- Extra ¼-inch cables.
- Headphones (optional, if they have a closed-back pair that they love).

Preparing various musicians for the studio
WORKING WITH DRUMMERS

One of the most important things that drummers often overlook is the importance of new drum heads. Having a fresh set of drum heads will give the drums more tone and sustain, especially on the toms.

The ability to tune drums becomes extremely important in the studio. Tuning can make the difference between drums that sound full and professional or drums that sound dead and lifeless. Poorly tuned drums can lead to a lack of differentiation between the tones of the toms. Having a drum tuner will help measure the tension of the drums around their various tuning lugs. There are many different brands and they sell for around $60, which is a small investment for a nicely tuned drum set. It is also a good idea to invest in one of these drum tuners to keep around the studio for instances where the inexperienced drummer may need help tuning his or her drums.

Drummer's headphones

Drummers, more than any other musician, are particular about what they hear in the studio. Because drums have a much louder sound to them, the drummer may not be able to hear the different instruments in the headphones above the acoustic sound in the room. Most headphones that drummers use in the studio are completely isolating to cut down as much of the external acoustics as possible. This makes it crucial to give the drummer a dedicated cue mix when recording. It can be a good idea to select headphones that the drummer is familiar with using as long as those headphones will block out most of the sound in the room.

WORKING WITH GUITARISTS

Every musician is very particular when it comes to his or her equipment, and a guitar player is no different. Most guitar players in a band will have a specific guitar rig tailored to their style of music. If the guitar player has a couple of different amplifiers, it may be a good idea to bring them into the studio so that you can dial in different guitar tones for the different parts.

Guitar effects processors

A typical guitar rig can consist of a combo amp or a preamp, power amp and perhaps some sort of effects processor or pedals. Effects processors and pedals are where you can run into some trouble. The $200 digital effects processor purchased in the mid-1990s is going to contribute substandard-quality effects to the sound. This is where you, as an engineer/producer, will have to make the call as to whether these effects are contributing to the sound of the instrument or are creating problems that cannot be solved in a mix.

Certain effects can easily be created in a digital audio workstation, such as chorusing, flanging, echo, etc., however, there are some classic sounds that come from pedals, amplifiers, and quality digital effects processors. This is where you need to work with the guitar player, in advance of the recording session, to determine which route he or she wishes to go. Committing to a sound to "tape" can be advantageous in some regards, but this can also present limitations later on when the musician may wish to remove effects or processing in the mix but cannot.

Preparing a guitarist for the studio

Make sure that the guitar player is headed into the studio with several sets of new guitar strings. Nothing will kill the flow of a recording session more than having to head to the local music store to pick up new strings, and good luck to you if you are recording at 8 PM or on a Sunday and need to replace guitar strings.

For days of intensive guitar recording, it can be helpful to change the guitar strings each day. This prevents an overall change in guitar tone from parts that were recorded on fresher strings that were overdubbed with duller strings.

Having a guitar "set up" by a professional, dedicated guitar shop can help eliminate problems with intonation before the recording session even starts and will help the band make the most out of their studio time.

Good-quality guitar cabling is also advantageous. It is always good to have nice cables in the studio to prepare for the forgetful musicians, or those who stroll in with their road-worn equipment. A good-quality cable will help shield unnecessary radio frequency (RF) noise from being picked up in the recording.

Audio cables should always be laid out separately from the AC power cables, as the placement of these cables alongside each other can bring unwanted interference into the audio.

WORKING WITH BASSISTS

Having a bass player put fresh strings on his or her instrument before heading into the studio will have the same benefits as a guitar player with fresh strings. Bass strings, in general, will take longer to lose some of their clarity over time, but having a new set of strings will help the bass cut through the mix.

Again, this is the time to make the decision based on what is best for the song. Some classic basses will have a rounder low end, with not as much high-frequency presence as their more modern successors. A more modern bass may work better with newer strings than an older bass.

Sometimes bass players will only show up to the studio expecting to play through a direct box. Some of the best bass tones recorded have been done only with a direct box, so knowing in advance whether a bass player is planning on showing up with or without a bass amplifier will help you choose how to arrange the band in the studio. When going though a direct box, it is common to place the bass player—and very often, the keyboard player—in the control room. If eye contact is important, and the bass player is only using a direct box, then it is easy to place the bass player in the same room as the drummer.

Bass amplifiers

There are almost as many different bass amplifiers out there as there are guitar amplifiers. A cheap bass amplifier may not contribute near as much to a quality recording as a high-end bass amplifier. Some engineers prefer to record the bass via a direct input (DI) box, and some prefer to record a microphone directly on the bass cabinet. Several bass combo amps and preamps will have a direct output on the back of the unit. A cheap bass will be subject to having a cheap-sounding direct output, so a quality DI box will be much more useful. Also, many of the direct outputs on these devices will create problems as their output may be colored by the equalizer (EQ) setting on the preamp that is set up for the acoustic tone of the instrument.

The most flexible thing that you can do in the studio is to record both a direct input, from a quality DI box, as well as from a microphone on the amplifier. Every engineer and style of music is different, and so the means to capture the bass can be varied, as there is no single right way to do it.

WORKING WITH VOCALISTS

A vocalist may not think that he or she needs to prepare as much for the studio as the other instrumentalists, but this is not the case. Vocalists should always bring a printed copy of the lyrics sheet. Even if the vocalist has the lyrics memorized, it is very helpful to the engineer to have the lyrics sheet available when punching in vocal lines.

If the singer chooses to keep a lyrics sheet in front of him or her while singing, it is helpful to have them typed, and on either one or two sheets of paper. Nothing will ruin a good vocal take like the sound of a paper rustling in front of the vocal microphone.

Coming into the studio with a selection of teas and throat sprays that the vocalist is familiar with using will be beneficial. Vocal overdub sessions can be very long and wearing on the singer, but having a nasty-tasting throat spray may not help anything at all.

CLICK TRACKS

Using a click track in the studio

Click tracks have been used in the studio for years, so much so that they are not considered a studio trick, but this has not always been the case. The origins of the click track originally came from the use of film and video in order to give musicians the precise timing for the musical cues that they were recording. This spread to the world of music production, so that the artists could record their parts with a solid rhythmic reference.

ADVANTAGES OF USING A CLICK TRACK

Since many of the techniques in this book are to correct timing as well as create sequenced parts for a song, having a click track becomes very important. This will allow the splicing of different takes, even across different tracks when editing the song together. This will also make it easier to add sequenced parts, as well as correct the timing during editing.

There are ways to create a click track to parts that are prerecorded. However, to give you the best-sounding, professional recording, starting with a click track is very important in most pop recording circumstances.

PUTTING TOGETHER A CLICK TRACK FOR THE DRUMMER

With as many different metronomes and drum machines that are out on the market, there are as many different click track sounds. It is best to stick with one sound through the duration of the recording. If the drummer is not used to playing with a click track, the entire session will be impacted. Find out if the drummer regularly rehearses to a click track. If the drummer is used to hearing a specific sound, try to build a click track to that sound or one very similar. If the drummer is used to rehearsing with a high-pitched "beep," then it makes sense to record to that same sound. If a less-experienced drummer suddenly comes in to record and has to record to a cowbell sound as opposed to the high-pitched beep that he or she is used to, the odds of having problems with the tempo during recording increases.

CHOOSING THE SUBDIVISIONS FOR THE CLICK TRACK

Depending on the song, sometimes drummers prefer to play to a prepro-grammed drum loop as opposed to a simple click track. This has the advantage of carrying the subdivisions that may be needed for particular songs. The slower the tempo of the song, the more important it is to have subdivisions in the click track to prevent the drummer's tempo from fluctuating too much in the middle of each bar. Along the same lines, having subdivisions in a song with a fast tempo can clutter up the spectrum in the drummer's headphones and make it more difficult to hear the individual beats.

CLICK TRACK FUNCTIONS OF VARIOUS DIGITAL AUDIO WORKSTATIONS

Each DAW will have its own click track functionality with varying types of clicks. The options for these click tracks can be very limiting, by having a small selection of different sounds and accents of the click track. Oftentimes there will only be functions to determine which beats to accent and whether or not to put subdivisions in the click track.

Creating a custom click track

Creating a custom click track for the song allows for much more flexibility with the click track. A software synthesizer with a modular drum machine can make for an excellent device for creating click tracks. The ability to select each individual sound of the click track, as well as creating the desired subdivisions, can become very handy when creating click tracks.

Simply creating a click track with one of these stand-alone devices and then exporting the sound as an audio file at the appropriate tempo can make a very useful click track when placed on a track in a DAW.

If you come up with a particular beat that your drummer finds comfortable playing with, then you can use that same click track, but just export the audio files for the various tempos.

Selecting the appropriate tones for a click track

It is important to carefully choose the tones you select for a click track so that the musicians can best hear them while all of the other instruments are playing. If you listen to a metronome or other dedicated click-generating device, the sounds are invariably very percussive, with little to no decay in the sound. These tones are also usually high in pitch. If you are building a custom click track for the band to follow, keep those same principals in your tone selection.

A higher-pitched sound will cut through all of the instruments better as opposed to a lower-pitched tone, which can blend in with acoustic sounds of the drum kit and other instruments. Having a specific sound that is not mimicked by any of the other instruments being recorded helps to differentiate between the click track and the sounds of the actual musical performance. This is why you would choose a cowbell type of sound, over that of a hi-hat cymbal. (Insert overused 'more cowbell' joke here).

The decay of a sound is also important. If you are trying to accentuate quarter notes, then it is best to have a sound that is percussive, and yet decays quickly, so as not to continue over into the other beats of the click track.

Using Reason to create a click track

The Redrum unit in Propellerhead's Reason is an excellent device to make click tracks (Figure 1.3). It has a total of ten sounds that you can add to a click track, with many different individual samples to choose from. In addition to the different samples, you can also decrease the decay of the sounds so that their envelope is short enough that the sound does not decay into the next beat. You can also adjust the pitch of the individual sounds to find appropriate room in the sonic spectrum for the individual sounds. Panning these sounds individually in the Redrum unit will help to differentiate the sounds in a headphone mix.

FIGURE 1.3
Reason's Redrum module programmed as a click track with subdivisions.

You can go about creating these click tracks in a couple of different ways. The first method to try would be to use Reason as a stand-alone device, and merely export the click track as a WAV file as long as you have the appropriate tempo. This WAV file can then be imported into a specified track in your DAW for the click track.

The other method of using Reason to create a click track is to use the rewire functionality of various DAWs to control the tempo in Reason. The advantage of using it this way is that you have the ability to create a single click track file in Reason. You can then open that same file in the different session files of your DAW, and Reason will adjust the tempo of the click track to match the tempo setting of your recording software. One thing to note, however, is that if you are using different time signatures in the different songs, it is helpful to create a separate click track that matches the feel of the different time signatures.

Rewiring Reason will allow you to have multiple outputs feeding into your DAW. This allows for the ability to have multiple click tracks. This can come in handy if the drummer wants a very subdivided click track and the rest of the band finds the extra beats distracting. You can then provide them with a simpler click track with a different Redrum unit with its outputs going out a separate channel via Rewire (Figure 1.4).

When setting up the host DAW for multiple click tracks, take note, as seen in Figure 1.5, that the Rewire plug-in has separate outputs set up for Reason. When you set up different outputs in your host application via Rewire, you then have to set up Reason to feed to those separate outputs from the different devices (Figure 1.6).

Tailoring the click track for drummers

Drummers are often the key to a successful recording with a click track. If you are working with a band that has little recording experience, and most of their performances are relegated to small clubs and their rehearsal space, odds are that the drummer has not rehearsed to a click track.

FIGURE 1.4
The Rewire plug-in displaying the output from Reason.

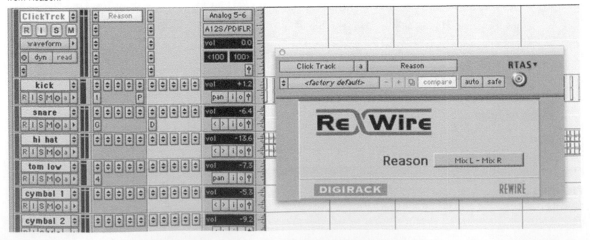

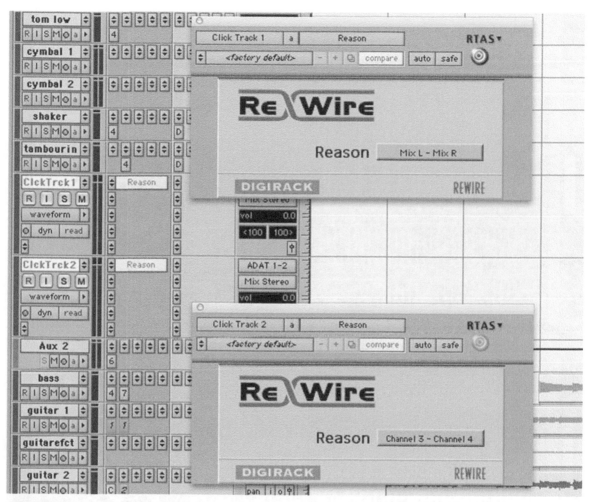

FIGURE 1.5
Multiple instances of Rewire with separate outputs from Reason.

Having a drummer who has not played to a click track before suddenly come in to the studio and try to play along with one can be disastrous. The drummer will fall behind or ahead of the click track, and then suddenly adjust his or her performance to be right on the click track. This creates a noticeable jump in timing during the performance. It winds up being worse than if you were not recording to a click track at all.

A technique to try with a drummer is to create an audio file consisting only of the click track with which the drummer can practice along. There is a brand of headphones called Metrophones™ that have a metronome for the drummer to rehearse with, as well as a line-in jack to plug an external source into the headphones. These are ideal headphones, as they can allow the drummer to rehearse to a certain tempo with the headphones, as well as rehearse to a prerecorded CD with the actual click track that the drummer will record to.

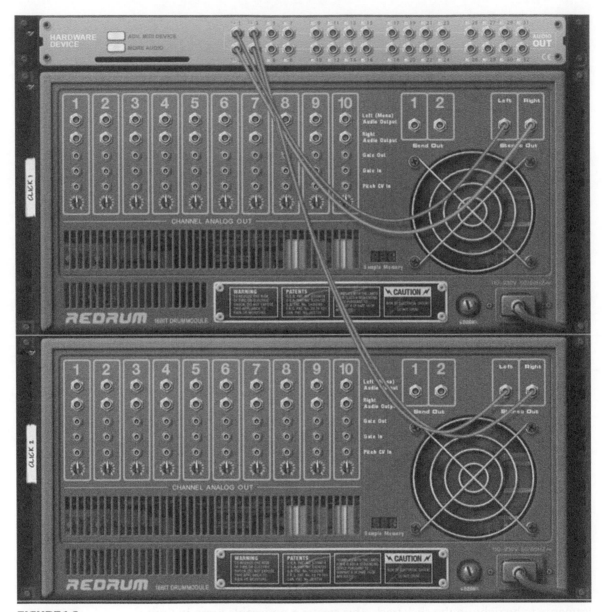

FIGURE 1.6
The outputs of the different devices in Reason assigned to different Rewire outputs.

A VARIATION ON THE STANDARD CLICK TRACK

A helpful variation on putting together a click track for the drummer is to have the other musicians in the band record scratch parts of their instruments directly to the click track, prior to recording drums. This will only work if you are intending to focus completely on the drums during basic tracking, and replacing the other instruments later. The reason for doing this is that the drummer can focus completely on staying in time with the click track without trying to follow a guitarist who may be rushing the click track.

This technique will also allow you to make a CD for the drummer to rehearse with that will contain the tracks that he or she will record along to, once in the studio. You can also optionally create a rehearsal CD that will have the instruments on one side and the click track on the other.

WHAT ELSE WILL A CLICK TRACK DO FOR ME?

Having a click track will enable the engineer to splice together several different takes of the same song relatively seamlessly. You may have a take where the intro and verse are great, but the chorus is shaky. You can then copy a different chorus in the same take of the song with relative ease. This will be covered in depth in the editing section, but oftentimes editing is done during the course of basic tracking.

Using the click track with other musicians

There are different methods when working to a click track with the other musicians other than drummers. There is no single right way to do it, only what works best for the artists. Creating a click track may take a bit of experimentation at the beginning of the session, so make sure that the musicians are aware of it.

Sometimes the other musicians will want to hear the click track that the drummer is hearing. If the drummer is solidly locked on to the click track, then this works very well. If the drummer's tempo is fluctuating slightly from the click track, then the musicians may accidentally choose to follow the click track, instead of the drummer, and this can distract the drummer, throwing his or her timing off, as well as lead to a sloppier performance of the basic tracking.

If the drummer's tempo fluctuates slightly from that of the click track, then it is best to remove the click track from the other musicians' headphones and have them lock into the performance of the drummer. Inexperienced drummers may tend to follow the musicians more than lead them, so in these situations you will have to figure on the click configuration that best suits the artists.

Headphone bleed when using a click track

One of the biggest dangers when using a click track is the "bleeding" of the click into the microphones. This bleed can become more apparent during quiet sections of the song or at the very end. With preplanned click tracks, you can set the click track to stop precisely at the very last hit of the drums. This prevents the capturing of the clicks in the microphones while the cymbals are decaying. If you are an attentive engineer, you can manually mute the click track precisely at the last beat.

A high-pitched click track will be more susceptible to headphone bleed than a low-pitched click track. Those with simple high-pitched beeps are the biggest culprits for being captured by microphones, but those are also the easiest to hear with all of the instruments playing at the same time.

The more headphones in the same room that have a click track coming through them, the greater the risk of capturing bleed into the microphones. With instruments that have a higher SPL output, such as a loud guitar amplifier, the danger of picking up bleed from the click track is much lower. This is due to the fact that the gain in the microphone preamp is much lower than it would be if you were recording acoustic guitar.

With a large choir, there can be several different headphones in the same room with a relatively low volume when compared to a drum kit. For these situations, it makes sense to use a low-pitched source for the click track, such as the sound of a kick drum with the high frequencies rolled off. The bleed from this type of click track is much less apparent and can potentially be eliminated from the recording through the use of a high-pass filter on the recorded tracks.

Automating the level of a click track can become handy for sections when the volume of the instruments drops substantially. An example of this would be during a break when the drums stop playing, but an acoustic guitar may continue. Headphone bleed is notorious in acoustic guitar tracks and during the decay of cymbals, so keep this in mind.

In order to detect click bleed in instruments, it is always helpful to periodically monitor the sound of the recording with the click track turned off in the control room. This will allow you to hear if there is any bleed coming in from the click track, and which microphones may be the culprit. When listening back to the various takes, in the control room, turn the click track off so that you can monitor the tracks for any click track bleed, as well as listen for obvious timing inconsistencies.

Variations to playing with a click track

PLAYING TO A CLICK BEFORE THE SONG EVEN STARTS

There may be times when jumping into the recording of a song using a click track can be tricky for inexperienced musicians. When recording, you may find that it can take a few bars for the drummer and band to settle into the click track. The tempo may take a bit of time to settle into the drummer's head. In instances such as this, it may be beneficial for the drummer to play something simple for four or eight bars before the actual start of the song to settle into the tempo. This will help the drummer lock into the tempo before the actual recording begins by playing a kick and hi-hat groove. If you are recording with this method, make sure that anything the drummer is recording prior to the beginning of the song is easily editable to create a clean actual start to the song.

ALTERNATIVES TO RECORDING WITH A CLICK TRACK

There are a few different reasons why you would choose not to record to a click track. If the drummer is actually performing worse to a click track than if he or she were not using a click track, then it stands to reason that using a click track will inhibit his or her performance and actually make the recorded track sound worse.

Sometimes you will find musicians that will choose not to record with a click track for artistic purposes. Perhaps they want the tempo to fluctuate slightly according to the band's performance. Sometimes rock musicians will choose this route. If you are recording traditional jazz, a click track is almost never used, so as to accommodate the feel of the performers.

There are a couple of different methods to help the consistency of a song, without recording to a click track.

STARTING WITH A CLICK TRACK

Sometimes musicians will feel hampered by a click track, but if they are solid players, the tempo suffers little fluctuations. In instances such as this, the click track can be used merely to start the drummer off. Playing the click track up until the downbeat of the song can help create consistency with the recorded tempo of the track. Additionally, if there are no major fluctuations, then there becomes an increased chance of splicing together different sections of the song from different takes to create a solid rhythm track.

REHEARSING TO A CLICK TRACK IN THE STUDIO

If the entire band is recording at the same time, it could be beneficial to rehearse the song for a few minutes with the click track, and then quickly jump into recording the whole band without the click track. This helps to lock all of the musicians into the tempo before the actual recording takes place.

Creating a click track and tempo map from an existing drum track

If you are not recording the drum tracks to a click track, but need to add one for the other musicians after the fact, there are software solutions that can do this. This can be accomplished by matching the tempo map in your DAW to the recorded performance.

The added benefit of creating this tempo map to the performance is that you can add sequenced parts and quantize them, not to a predefined tempo, but to the actual tempo of the performance, down to the bar or beat if you choose.

Using Beat Detective to generate a tempo map from a performance

Digidesign's Beat Detective, in addition to being used to correct the timing of different instruments, can also be utilized to create a tempo map from an existing performance.

DEFINING A STARTING SELECTION

To create this tempo map, it is best to begin by creating an initial tempo from the drum tracks. You can do this by selecting the percussive elements of the drum tracks that define the beat. Usually this is the kick and snare, but can include the toms or overheads.

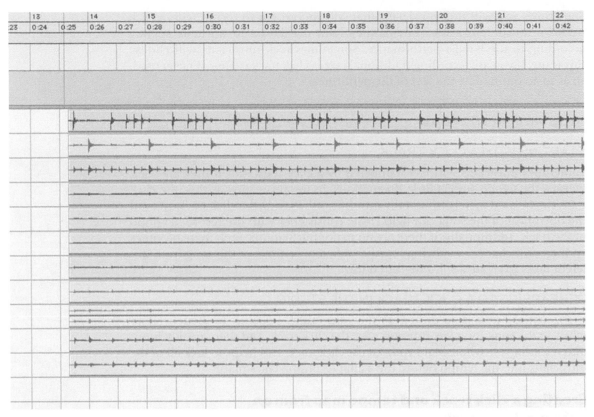

FIGURE 1.7
Recorded drum tracks with no associated tempo map.

In Figure 1.7, you can see several tracks of drums that are going to be used to create a tempo map. They will be used to create a click track for the other musicians, as well as sequenced loops that will be used throughout the song.

From here, the focus is on the kick and snare track as the source for the tempo map generation. You can see in the beats ruler at the top of the window that the actual drums start between bars 13 and 14, but the tempo map is going to be adjusted so that it will fit with the recorded tracks.

SELECTING THE BEGINNING BAR

It is not a good idea to start the recording audio right at the very beginning of the session counter. It can complicate issues if you are looking to create audio that may take place before the downbeat of the main instruments, or connect to an outside device through Rewire.

Creating a tempo map so that the main instruments start at bar 101 will help keep track of the bars by merely subtracting 100. If you create the tempo map so that it starts on bar 1, then every bar previous to that will be a negative number.

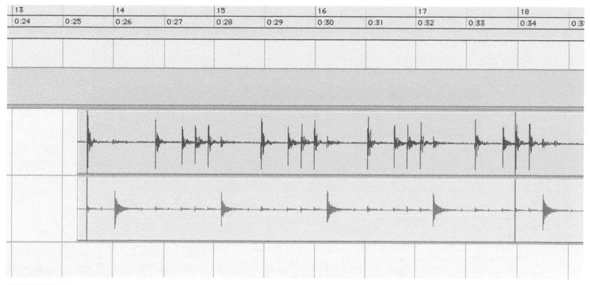

FIGURE 1.8
Separating four measures for an initial tempo map generation.

This can create problems if you are adding sequences from hosted external software, such as Reason, which will not sequence audio before bar 1.

The way to first begin to create a tempo map is to separate the regions by selecting a precise number of bars. In Figure 1.8, you will see that the kick and snare regions have been separated precisely by four measures. These will become bars 101–105 on the newly generated tempo map.

It usually works best to select a short region to begin with, and then perhaps increase the amount of bars you use to generate the tempo map. If Beat Detective easily calculates 4 or 8 bars correctly, then you can expand this to 16 or 32 bars. If the accuracy of the detection seems to suffer, drop down to 2 or 4 bars.

By selecting this exact 4-bar region and opening up Beat Detective, you can then tell Beat Detective that these are 4 bars by entering bar 101 as the start bar and 105 as the end bar (Figure 1.9).

Once you have this defined, you can click "Generate" and Beat Detective will adjust the tempo map for those 4 bars to create an average tempo that those 4 bars are creating. You can see in Figure 1.9 that Beat Detective has determined that the average tempo of this section is approximately 113.3 beats per minute.

The drummer's performance may fluctuate with respect to the different bars, and so using the analysis feature of Beat Detective will further allow you to create a more precise tempo map. This will make any loops added synchronize better with the drums.

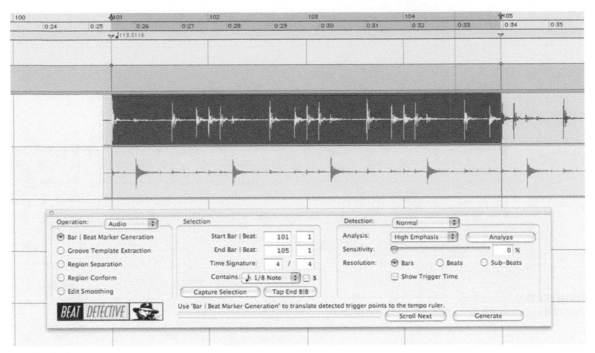

FIGURE 1.9
Four measures entered in Beat Detective, reflected in the tempo map.

ANALYZING THE AUDIO

Beat Detective will analyze the transients to more precisely determine when the actual beats are performed with regard to the tempo map. In order to accomplish this with optimal precision, determine what the shortest subdivision is in the beats and then select that under the "Contains" feature of Beat Detective, which is under the "Selection" portion of the window.

Moving to the "Detection" portion of the window, you have the choice under "Analysis" to select "High Emphasis," "Low Emphasis," or "Enhanced Resolution." In most situations, any one of these three choices will work equally as well as the other. You can choose "Low Emphasis" if the frequency content is in the lower range such as a kick drum, and "High Emphasis" if the content is in the upper-frequency range such as hi-hat cymbals. The safest choice is to select "Enhanced Resolution," as this has the newest transient detection algorithm.

Once you click "Analyze," the grayed-out sensitivity bar will become active, and from there you can raise it to see where Beat Detective has determined the beats to be. By selecting the resolution to be "Bars," the tempo map will be created with the fluctuations of the tempo being adjusted at the start of each bar. This is usually good enough for most loops and sequences, but you can further enhance the resolution by selecting either "Beats" or "Sub-Beats" in the selection window.

The "Sensitivity" bar should be raised up to where you see the Beat Detective place an indicator at the beginning of each bar, but no higher, as the markers can jump out of time depending on the content of the audio. From here, you can click "Generate" and Beat Detective will put appropriately timed markers at the beginning of each bar, so that the tempo map now follows the drums (Figure 1.10).

For these 4 bars, Beat Detective has created slight tempo adjustments to the drummer's performance. The tempo previous to this point has been adjusted, and now the tempo map prior to the section has been adjusted as well.

To complete the song, you can utilize Beat Detective in this same manner, advancing 4–8 bars at a time. If there is a stop in the drum part, you can merely count out the bars. Highlight these bars with the selector tool, enter their length into Beat Detective, and then click "Generate." This will calculate the tempo based on these starts and stops despite there being no audio present in the drum tracks.

If there are drum fills, you may need to select those single or double measures individually so that Beat Detective is not trying to analyze a section that is too long. Additionally, you may need to listen and compare these sections for accuracy.

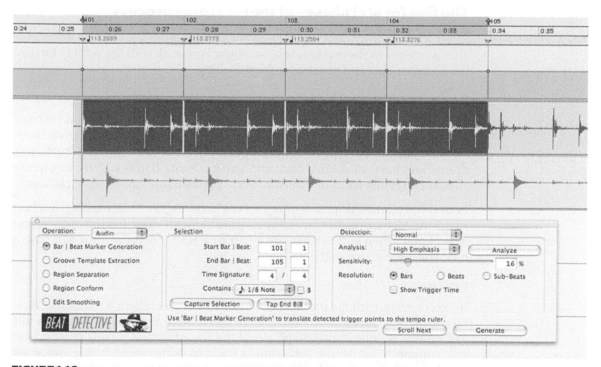

FIGURE 1.10
Four bars, with the tempo map reflecting the slight variations of the drummer's performance.

CHECKING THE ACCURACY OF THE TEMPO MAP

One method of checking the accuracy of whether you have correctly generated the tempo using Beat Detective is to create a single track and add the "Click" plug-in. This plug-in will follow the newly generated tempo map and you can listen to hear if the tempo map is accurate.

THE RECORDING PROCESS

Rather than go into depth as to the actual recording process (there are many great books on the market that successfully achieve this), we will go in to some of the techniques used when actually working with musicians, and some of the tricks that professionals use to make a recording session go smoothly.

The engineer/producer

When you are engineering a record in a home-based studio, or with clients who do not hire a producer, oftentimes you become a coproducer for these musicians. While they are recording, it is partially your responsibility to identify potential problems with the performance. This is especially important when working with musicians who have little or no studio experience. It becomes your job to educate them as to the process, so they know what is going on for the entire time.

The five (or six) phases of the recording process

PREPRODUCTION

This is covered in the previous section, where you work with the artist to finalize the arrangements and prepare for the recording.

BASIC TRACKING

This is the main recording portion of the process. This is when the foundation of the song is being captured. All of the other parts of the song are going to be built on what is captured here. Basic tracking can consist of just capturing the drums; drums and bass; or drums, bass, and guitars. Some bands will try to capture almost everything short of a finished vocal performance during basic tracking.

Even if you plan on disposing of certain tracks during the basic tracking, it is best to do the highest-quality recording as possible with these instruments, as you never know when you will capture a part or a sound that the artist will have a hard time replicating in the future.

EDITING

In recent years, with the proliferation of DAWs, there is also a good deal of digital editing that can be done. Even though this is listed as a separate phase, in this age, editing is done during all phases of the recording and mixing process. You will constantly be editing parts during basic tracking, overdubbing, and even mixdown, so there is rarely a time when you are going to sit down and take care of all the necessary editing in one sitting.

OVERDUBS

This is where the rest of the song is put together. Vocals, guitars, keyboards, percussions, MIDI tracks, etc. can all be added during this phase.

MIXDOWN

This is often considered the most important phase of the recording. These days, mixers of major record labels oftentimes get almost as much of the credit for a record as the producer does. When working with a band, make sure that their budget does not fall short when it comes to the mixdown process, as this is the point when you can turn some tracks into a final professional product. This phase can make the difference between a recording sounding like a demo or a high-quality product for a major label.

MASTERING

This phase will be covered more extensively toward the end of this book, however, it should be stated that this is also not an area for a band to cut costs. A quality mastering job will make a mix and make it jump out of the speakers at home or on the radio on par with other professional-quality recordings.

What makes a successful engineer?

Being a recording engineer requires skill beyond knowledge of equipment, acoustics, technology, and an ear for making records. It also requires skills in working with clients, as well as on-the-fly troubleshooting skills.

INTERPERSONAL SKILLS

Whether you are a home studio owner or working in a major studio, learning how to work with clients and make them feel comfortable is one of the most important skills that you can possess. This is what makes a classic producer. Clients of a studio are often under stress to get their recording done. There are budget concerns and performance concerns. Having a laid-back and adaptable personality will make you much more successful in the long term, than if you were to contribute to the tension of a recording session.

For the most part, studios need to accommodate many different styles of music and a wide variety of clientele who come into the studio. You may find yourself recording rock, jazz, hip-hop, country, Spanish, reggae, folk, and any other style of music that is out there. With these different styles comes a different assortment of people, but their main goal is always the same, which is to make the best-sounding record possible.

To be able to work with all of these diverse people is a great strength for the recording engineer who is looking to build up repeat clientele. In smaller markets you will need to be able to work with just about anybody who can afford to walk in your door.

MUSICAL KNOWLEDGE

Throughout the history of recorded music, recording engineers come from a wide variety of backgrounds. You will find a varied background when it comes to the musical knowledge of these engineers. Some are virtuosos on many instruments, and others have an electrical engineering background.

The helpfulness of a strong musical background cannot be understated. As mentioned earlier about the need to be a coproducer in a lot of situations, you may be called up to follow a written chart to appropriately punch in and out during basic tracking and overdub sessions. You may be called to help provide MIDI supplemental tracks to a recording. Here is where a musical background and some experience with arranging will come in handy.

If you do not possess a strong musical background, this does not necessarily mean that you will be any less of an engineer. There are certain musical skills that you should pick up to be successful. You should be able to count bars of music, as well as identify sections of a song. Be able to jump to the prechorus when the band asks, or cut the length of the bridge in half. Developing a musical ear can be just as important as a musical background.

Knowledge of musical styles

It is important to study and be sensitive to the musical style that you are recording. Not everybody's personal musical tastes extend into the style of music that comes into the recording studio. Since you will most likely have to cater to all styles of music, studying the sounds on the various recordings of that genre will help you come across as a professional engineer.

Of all the different styles of music that are out there, there are different techniques to record each of them. You will need to decide, along with the musicians, the best way to record them. The standard pop music recording method will cover most styles, including country, folk, reggae, etc. This is the standard basic tracking with overdubs and mixdown. Most of the parts are overdubbed and not captured during basic tracking.

Other live styles of music, such as classical, bluegrass, and jazz, require that most, if not all, of the tracks are captured live.

ROCK

No other musical style has pushed innovation in sound recording more than rock and roll. From The Beatles to Radiohead, the experimentation never seems to end. Rock music has so many different styles and variations that it is impossible to come up with a single method for capturing its style.

Raw rock music

If the band is looking for a rawer sound, then you are looking to capture the band's performance together. This becomes more important than isolation. For basic tracking with this sound in mind, you may be looking to keep drums, bass, and rhythm guitars.

For a more produced sound, the goal becomes capturing a solid performance from the drums and bass, with the idea of recording all the other parts during the overdubbing phase. Planning the studio plot with this in mind should be as much isolation of the instruments as possible.

JAZZ

For recording jazz, there are two main methods of capturing this in the studio, depending on the type of jazz.

Traditional jazz

For the more traditional jazz, which is performed live, and which the genre is more sympathetic to instrument bleed, the goal is to set up the recording space for the musicians to capture their natural performance live. This means there will be sacrifices to the isolation of the sounds in order to allow the musicians more eye contact with each other. If you have the means to put any lead instruments such as saxophone or trumpet in an isolation booth, that will give you the flexibility of overdubbing a part that has a few wrong notes. Even though you may have all the musicians in the same room, this does not mean that overdubbing a missed part is out of the question. With a well-laid-out room setup, isolation can be maximized by microphone placement.

The musicianship of traditional jazz musicians relies on individual dynamics, and so the engineer's job is to capture the performance of the musicians. These musicians do not want the engineer to control the sound in such a way that the technique and musicianship is lost.

Contemporary jazz

Creating a contemporary jazz sound is recorded in much the same way as a pop record is recorded. The sounds and performances are much more produced. You can look for a basic tracking session that may consist of all of the instruments, but isolation should be taken into consideration as well.

RECORDING THE BASIC TRACKS

The methods for recording basic tracks are covered exhaustively in many audio textbooks. Rather than go in depth and discuss what it takes to capture basic tracking, here are some tips that you will find useful when setting up the basic tracking.

The initial setup

Allow yourself plenty of time in the studio to set up microphones and headphones. Certain studios allow for a certain amount of time to be comped to the artist for setup. Whether you and the studio choose to do this or not is up to you, but be sure that the band understands at what exact point in time they are going to be billed for the studio time. There is also a fair amount of troubleshooting that takes place during the initial setup, so budget time in there as well.

When it comes time to set up for basic tracking, allow yourself an appropriate amount of time to set up before the musicians even get there. There are many things that you can do before they even arrive. You can set up all the anticipated microphones as well as the headphones for everybody. If you have met with the artists in advance, you will have already discussed the layout of the room and where all the musicians will be located.

Take time to make sure that all the equipment you have set up is working correctly. It is helpful to have more than one person present during a recording session: one person working as the engineer and another person working as the assistant engineer. This makes it easier for troubleshooting in the studio. You can have the assistant engineer test each of the microphones by scratching on them or speaking in the room and the engineer can listen to each of the microphones and make sure the signal coming through each microphone is clean. In addition to testing the microphones, it is also helpful to test the headphones and make sure that they are working properly.

An assistant engineer will also be able to help with microphone adjustments when you are dialing in the tones for each of the tracks. Having an assistant engineer will give a more professional impression to the artists as well as help move things along in a timely fashion.

If possible, have the drummer arrive at the studio before the other musicians, as the drummer takes the longest to set up. During basic tracking drums are the most important instruments to capture. Having too many musicians around can slow things down when you're trying to set up for recording.

Dialing in sounds

There is no definite amount of time that you should spend on setting up and dialing in sounds for the recording. You have to keep in mind how much of the time should be budgeted for setting up the tones with what needs to be accomplished in the time that the artist can afford. It takes time to dial in the EQ settings, and any other dynamic processing that you may wish to use. The stories of major-label artists spending days to get the exact snare drum sound they want are legendary. This is because the record labels are paying for months of studio time, as opposed to perhaps a few days for a local artist.

To print, or not to print (processing during recording)

Printing effects means to actually record the signal processing, destructively, to the track. This is oftentimes used when recording vocals. The engineer can use a compressor in the analog signal path to control the level and increase the signal-to-noise ratio of the recorded track. When recording to a DAW, these effects can be inserted after the track has been recorded, which would be considered nondestructive.

If you have an expensive-quality compressor that you enjoy using during recording, then by all means, print it. If you feel as though the bass could use some compression during the recording process, but only have a cheap compressor, then you may want to hold off on that compression and perhaps use a plug-in compressor instead, while recording. Equalizing should be done as a means to enhance or clean up the tracks. A simple high-pass filter applied to drum over-heads, guitars, vocals, etc. can really help to clean up the tracks and will not adversely affect the mixdown process. The tracks should sound great individu-ally, but focusing on how they sound altogether is much more important, so keep this in mind.

Setting up headphone mixes for the musicians

During the initial setup, be sure you are planning the headphone situation for all the musicians at the same time. The more individual headphone mixes you can give the musicians, the better off you will be in the long run. Most everybody wants to hear something a little different in his or her headphones, so trying to come up with a single mix that everybody can play along with can be a daunt-ing task.

CREATING A HEADPHONE MIX WITH YOUR DIGITAL AUDIO WORKSTATION

Since the proliferation of small and home-based studios, there have been changes in the way that headphone mixes are created. In a traditional studio, a headphone mix can be generated from the console by using auxiliary sends from the board, or by sending the musicians a duplicate signal of the stereo mix that the engineer is hearing.

Some vocalists will prefer to hear effects such as reverb or echo while they are recording. This can actually have an impact on the quality of their performance. One of the techniques that vocalists may want to use is to pan their previously recorded vocals to one side while they are doubling their part.

The DAW functions similarly to a console, in that there can be many separate mixes sent from the audio interface to the musicians. The amount of headphone mixes from DAWs is limited only by the amount of available auxiliary sends from the DAW and the analog or digital outputs from your hardware. There are a couple of different ways to set up these auxiliary sends internally from the DAW, but the most straightforward method is to set up a stereo auxiliary send on each channel. Then send the auxiliary out a pair of outputs on your computer's audio hardware. Then, connect these outputs to a set, either a headphone amplifier or a dedicated headphone system.

CONTEMPORARY HEADPHONE DISTRIBUTION SYSTEMS

Along with home studios and small digital studios, there are new and inventive ways to create headphone mixes for the musicians. There are stand-alone head-phone boxes that allow for musicians to dial in the mix that they want to hear in their own headphones. This can save a lot of time in the studio when you are trying to make everybody happy with their headphone mix, and your time could be better spent working on getting the best-quality tones for recording as possible. That is not to say that a good headphone mix is not important; it just allows for musicians to be 100 percent satisfied with their mix.

Aviom, Hear Technologies, and Furman each have a headphone distribution system that allows musicians to dial in their own mix. Each of these different headphone distribution boxes receives the audio signal via an Ethernet cable. This allows for convenient distribution with only having one cable to carry all the audio information.

These boxes can be set up so they can receive the stereo mix you are working on in the studio on a stereo channel. You can then route the other individual musicians, and even click track to the other channels so that the musicians can turn up the knob and hear more of themselves or others.

TIPS FOR SETTING UP A HEADPHONE MIX

Oftentimes, musicians are always wanting to hear "more" of something. Soon you find yourself turning every fader or send up to the point where there is no room to bring up what they want to hear. Rather than constantly bringing certain levels up, periodically ask them if there is something they would like to hear less of in the headphones to make their mix sound better.

When to start the actual recording

After you have all the microphones set up and the sounds dialed in the way you want to hear them, it is time to start recording takes. When the musicians are trying to record in the studio, it may take some time before they are comfortable in their surroundings. It can be helpful to start recording on an easy song, rather than a more difficult song. This will help boost their confidence and make it seem more like a rehearsal.

If the musicians are wanting to just run through the track, always record it, even if the musicians do not know that you're actually recording. Sometimes they may perform more relaxed, not knowing that you are recording the tracks. You may surprise yourself and the musicians by capturing a great take when they are just rehearsing.

In the days of analog recording, you were always concerned by the cost of analog tape, which created concern with which takes to record over. In the era of hard-disk recording, these concerns have vanished. You can keep dozens of takes on a single hard drive, without having to worry about how much more recording you can do, as today's hard drives have extremely large capacities.

Always take good notes when you are recording basic tracks. You may find yourself splicing together a single solid basic track from several takes. You may find that one take is good in the chorus while another take is good in the verse. Maybe you have a take where the drummer only liked his drum fills and not the feel of the rest of the song. Taking good notes will speed up the process of creating a composite basic track.

After each take that the musicians like, have them come into the studio and listen to it. This allows them to hear any trouble spots that may have appeared as well as give them a brief break from playing.

If the band seems to be stuck on a particular song, have them move on to a new song, rather than have their frustration build. They can then come back to the previous troublesome song once they are more relaxed.

If you need to edit different basic tracks together to create a single composite basic track, make sure you complete this before moving on to the overdubbing phase. Any timing corrections that need to be done should be completed before the overdubbing phase as well.

At the end of basic tracking, be sure you leave enough time for you to make the band a CD of the basic tracks they have recorded that day. The purpose of a rough mix is to give the band a reference for the tracks that have been recorded. The balance of the tracks should present all of the recorded material in a listenable format. This will help them to decide if the basic tracks they have recorded are good enough to begin the overdubbing process. It will also give them a reference with which they can rehearse and write parts to. Do not try to create a mix with any specific effects or unnecessary automation, as you are just trying to give a reference of the quality of the recorded material without any distractions.

Tips for recording individual instruments

TIPS FOR RECORDING DRUMS

When recording drums, there are usually several microphones placed. The more microphones that are set up, the more time it will take to mix the drums. Adding an additional spot microphone on the ride cymbal can be a useful option in mixdown. Since the ride cymbal functions counter to the hi-hat, it is good to have some additional control over the level in the mix.

Place a microphone specifically to capture the sound of the drums in the room. This can be used or discarded during mixdown. It can be handy to create interesting drum sounds, and it can also be gated and triggered with a snare drum for additional ambience.

TIPS FOR RECORDING BASS

When recording bass for a pop song, the bass player can easily be recorded in the control room, if the sound is just being captured from a direct input. It can be helpful to have a musician in the control room to give the engineer feedback as to how the song performed musically while recording.

Since most of the pitch center for the song comes from the bass, it can be helpful to always have the bass signal going into a tuner. This allows for the constant monitoring of the pitch of the bass. Since sending the signal from the bass into a tuner and then into a direct box can potentially contribute unwanted noise to the recording, it is best to create a send from the recorder into the tuner.

TIPS FOR RECORDING GUITARS

When recording guitars, it can be very helpful to record a direct signal through a DI box in addition to one or more microphones on the guitar cabinet. This gives you the flexibility, in the future, to send a direct signal to a different guitar cabinet, or to send the direct signal to a guitar cabinet emulator. This can either replace the sound entirely, layer different guitar sounds with the same performance, or help create a fake stereo sound.

There are two types of DI boxes used to capture direct input from instruments—passive and active boxes. In general, the active boxes will sound better, but there are some very nice high-end passive DI boxes as well. To use a DI box with the guitar, plug the guitar into the DI box and take the thru signal from the DI box into the guitar player's rig. The DI box will have an XLR output that can be sent to a microphone preamp and recorded like any other track.

When recording rhythm guitar parts, record an additional doubled version of the guitar part so that the tracks can be hard panned in the mix to make room in the center of the stereo spectrum for vocals.

TIPS FOR RECORDING KEYBOARDS

If the band that you are recording has a keyboard player, pay close attention to the keyboard sounds that the band is using. Many stock keyboard patches have effects on them that you may not want in the recording. Many of them will have reverb on their patches to make them sound bigger and wider. These reverb effects, found in keyboards, are not as good as the reverbs that will be used in the studio, so you may need to work with the keyboard player and get him or her to turn the reverb off on each of the sounds that he or she is using.

You may need to reference the manual of the keyboard in order to find out how to turn the effects off. Getting the keyboard player to bring the manual into the studio, or program the sounds to have no reverb before going into the studio, will save time. As with a guitar rig, some effects that come from a keyboard contribute to the sound, so make the determination, prior to recording, whether these effects contribute to the quality of the specific keyboard patch.

TIPS FOR RECORDING VOCALS

A pop filter should be placed in front of the vocal microphone, in order to prevent unnecessary pops in the recording from plosive consonants. There are different types of pop filters available. Some have a cloth that is stretched across a ring. Others are made from metal with slots that redirect the wind away from the microphone.

Be sure the gain from the microphone preamplifier has enough headroom in case the vocalist suddenly gets louder than expected. It is good to have the vocalist sing the loudest part of the song before recording to set the preamplifier accordingly; however, once all of the tracks are playing and the vocalist is singing, the vocal part can ultimately be louder than originally tested.

When recording vocals, it is important to keep in mind the proximity effect of directional microphones. As the vocalist gets closer to the microphone, there is a substantial increase in low frequencies. The engineer can use the proximity effect to his or her advantage to increase or decrease the low frequencies in the vocals by placing the vocalist closer or farther from the microphone.

Overdubbing

When it comes time to overdub parts, strategize the order in which you were going to record the different instruments. If you're looking to build a song from the ground up, start with the rhythm instruments. These can include bass, rhythm guitars, or keyboards. Focus on one instrument at a time. Many musicians will think that they are going to save time in the studio by recording more than one instrument at a single time, but it really takes more time because you have to stop and start more often, and it is harder to focus on the parts being recorded.

Editing during basic tracking, overdubs, and mixdown

The editing phase in the contemporary studio becomes crucial. Editing can take place during the basic tracking when you're splicing the different takes together to make a composite basic track. Running different processes such as Beat Detective to tighten up the drum parts also takes place during basic tracking.

Editing during the overdubbing phase consists of slight adjustments to the timing of the performances so they match the basic tracks. When you are overdubbing vocals, you may find that you have several different takes of a single vocal part. You need to take the time and carefully edit these different takes into a composite vocal take. The same holds true for all the other instruments recorded during the overdubbing phase.

When you are mixing, editing is done to clean up the tracks. To have the cleanest mix possible, it is best to eliminate anything that is not necessary in the mix. You can go through and clean up the spaces between the tom hits to eliminate as much background bleed as possible. Cleaning up any extraneous breaths from the vocal track will help out as well. With electric guitars, there is often no way when the guitar is not playing.

Recording MIDI tracks during basic tracking

If any of the instruments that you are recording has the capability of outputting MIDI data, then by all means you should try and capture the MIDI data simultaneously. This means including an MIDI interface as part of your studio rig. You should know how to capture MIDI tracks as well as audio tracks when you are recording. Today's engineer needs to have a solid understanding of MIDI in order to help arrange and supplement songs.

Capturing MIDI will allow you to send a keyboard player's performance to any other MIDI device in the future. Sometimes you may find that the sound that he or she is recording in the studio is not the best sound for the track. Having a selection of software instruments will allow you to change the keyboard sounds at any time in the future. This is especially true if the keyboard player has limited equipment.

MIDI tracks should be recorded in the same way that you are recording the audio tracks. That way when you edit the basic tracks together, the MIDI tracks will follow. Most MIDI tracks are added during the overdub phase, but if they are present during the basic tracking, it gives you the ability to quantize and edit the MIDI data in the future.

Eliminating elements

To route the recording process you'll find that you have recorded more tracks than you can use—this is not uncommon. There are many pop records out there that have over 100 tracks. There are no hard-and-fast rules as to how many tracks are too many. The biggest concern is whether the tracks are right for the feel of the song. You just want to make sure that you are not detracting from the song with the elements that are in the recording.

Just because you recorded a track does not mean that you have to use it. A song should not feel cluttered unless that is your goal. The main criterion for determining whether or not to use a track in the song is to decide if that given element helps or hurts the song.

Sometimes if you hear a song with all the tracks over a long period of time you will get used to hearing everything in there. It may help to have a fresh pair of ears to determine whether or not to keep a particular track.

Musicians are sometimes married to every single part that they put into a song. It takes a delicate touch from the engineer or the producer to convince the musician that that part is not necessary for the song. Careful planning of overdubs in advance will help prevent a runaway track count.

When you are recording it may be helpful to eliminate unnecessary parts as you record them. Determine whether or not they are contributing to the song as the musicians are recording them. This will help prevent the musicians from getting used to hearing every overdub in the future before you decide that the part is unnecessary.

GENERAL DIGITAL EDITING

There are many different types of editing that can be done to tighten up and enhance your project. Editing can be done to clean up noise, tighten up performances, and rearrange the song. In general, there is not a single editing phase, as it takes place over the course of the entire recording project. If you are doing basic tracking and overdubbing, the editing that is usually done is to tighten up the parts to make sure that you have a clean performance. Editing can also be done to rearrange the song. This is usually done during the basic tracking phase, before any overdubs have been added. In mixdown, you can further clean up the tracks to eliminate unnecessary noises from the tracks.

Editing throughout the entire process must be done carefully so that you do not generate any unwanted artifacts. Most digital audio workstations have different modes of editing. There is a general mode for editing that functions like a word processor where you copy, paste, and cut. This is the mode of editing that is employed most often.

Cleaning up the tracks

The first type of editing that you may do is to clean up the recording. When you start and stop recording there are oftentimes spaces before and after the instrument's performances. These can leave unwanted noises that may be heard in the mix. Simply removing these sections can help tidy up your tracks.

Background noise in an audio recording can have a cumulative effect. Multiple tracks with background noise will sum together to possibly create noticeable noise. When recording, there are oftentimes tracks that have just a few seconds of audio, but when recorded from beginning to end, they will have background noise for the remainder of the recording (Figure 1.11).

When you are doing these edits it helps to know the different shortcuts in your DAW. Being able to quickly zoom in close to waveforms will make cleaning up your tracks go much faster. When learning shortcuts, do not try to memorize them all at once. It may be a good idea to keep a printed list of all the

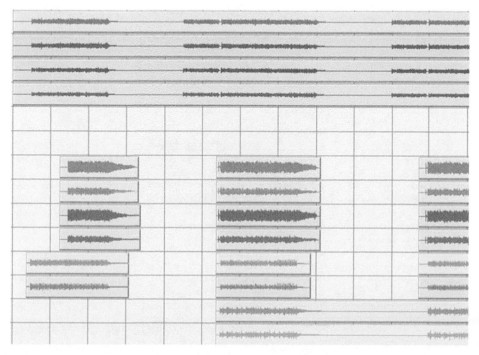

FIGURE 1.11
Multiple tracks of audio that can be cleaned to eliminate extra noise.

keyboard shortcuts when you are editing. If you are trying to memorize the shortcuts, start with just a couple of the most-used key commands that you think you might use. Zooming in and out horizontally will be some of the handiest shortcuts to know in the beginning. Pro Tools allows for single-key shortcuts if they are enabled. These shortcuts will really help you maneuver around your session.

When cleaning up the beginning of audio tracks, you do not need to concern yourself with the decay of instruments as you would at the end of the performances. Simply doing a hard edit may be all that is necessary to clean up the beginning of the parts.

When you are working with electric guitars, oftentimes the sound of an electric guitar will fade into noise that is coming from the amplifier. Editing these parts by applying an additional fade during the decay of the guitar will help eliminate the noise from the recording (Figure 1.12).

An added bonus to eliminating unwanted regions of a track is that it will reduce the amount of data coming from your hard drive during playback and recording. Sometimes too many edits, close together, can create more data coming from your hard drive than your system is capable of handling. You will find this if you do extensive editing using Digidesign's Beat Detective.

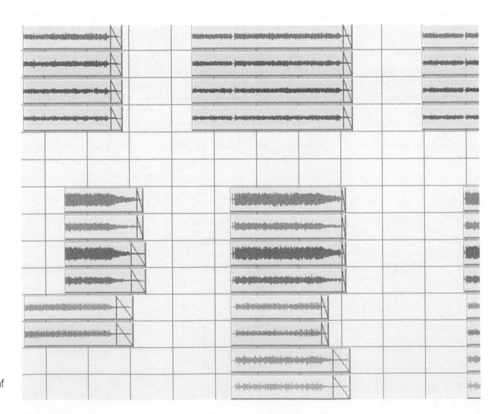

FIGURE 1.12
The same tracks as in Figure 1.11, with the empty spaces cleaned and fades at the ends of regions.

Editing toms

When mixing from analog tape, there are a couple of different ways to eliminate noise from the toms. The engineer could either run those tracks through a gate, which could be tricky with a sloppy drummer, or he or she could manually turn the toms on and off through the careful use of automation. Gates are used in recording to eliminate any unwanted noise by reducing or eliminating the audio until it crosses a specific threshold. It then opens up the audio signal until the level crosses below the same threshold. Using a gate is not always 100 percent accurate. Using the automation to turn on and off the tom tracks can be very tedious and does not allow for the tom tracks to decay naturally. They are either on or off.

FIGURE 1.13
Tom tracks unedited.

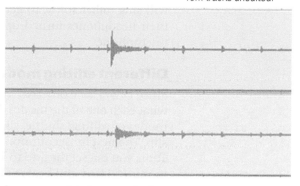

Sometimes you may want to hear the toms louder in the mix, but the amount of background noise in the tracks makes it difficult to bring up the toms without bringing up the instruments that bleed into that track. Manual editing of the tom tracks in a DAW allows you to accurately hear the toms only when they are playing, as well as create decay into silence through the appropriate use of fades.

You still need to be careful when you are editing the tom tracks, as sometimes the waveform displayed in your DAW can be misleading. There can be a snare hit in a track that appears louder than the toms are. It is still best to listen while you are editing to make sure that you do not miss any of these hits. If you are using Pro Tools as your DAW, duplicating the playlist before applying these edits will give you a means of going back to the original unedited part.

As with any edits, be sure to listen to them to make sure that you are not cutting off a noticeable amount of decay, if you are applying a fade (Figure 1.14). It does not hurt to make the fade last longer than the decay of the tom. If the fade is too quick, you may hear the background noise in the tom track abruptly disappear in the mix, which makes for a noticeable artifact. Sometimes applying a fade to take place during another strong hit in the drums will help mask any audible decay in the tom track.

FIGURE 1.14
The toms from Figure 1.13 edited and faded.

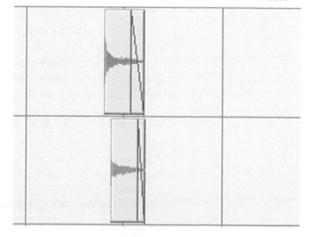

Editing vocals

When you are editing vocal tracks, there is some artistic license in deciding what to keep in the track and what to remove. Sometimes you may want a vocal part without any breaths. Sometimes you may want to keep the breaths in the vocal track to make it sound more like a natural performance. It is common to hear a lip smack before singing. Since vocalists are so close to the microphone when they are recording, a microphone will pick up any sound that the vocalist makes before, after, and during the overdubs. Adding compression of the vocals will only exaggerate these artifacts.

When editing vocal tracks, there can be many different takes from the vocalist. These takes are edited into one or more "comp" tracks. These can be edited into a single track or separate tracks for the verse and choruses.

Cleaning the vocal tracks when the vocalist is not singing is important because oftentimes you will hear lyrics pages turning, the singer singing along to the track, or any number of unwanted interesting sounds. Sometimes vocalists will sing rather softly, to get their pitch prior to coming in. If the singer needs their headphones turned up loud, then that bleed will be picked up from the microphone.

Different editing modes

Since DAWs have different modes of editing, you will find it very useful to know what each one of the modes does in depth. Digidesign's Pro Tools has four main modes of editing: shuffle, slip, spot, and grid. The most basic mode of editing is slip. The next most common mode of editing in audio production is grid. In Pro Tools you can set the grid to be one of many different types of units. It can be set to musical bars and beats or units of time such as minutes and seconds.

EDITING IN GRID MODE

Setting the grid to be measures of the song allows you to rearrange the entire song. This will only work if the instruments were performed to a click track that is matched to the tempo map of the session, and there are no tempo changes in the sections that you want to rearrange. Suppose you have a great rhythm guitar performance in the first chorus of the song, but the second chorus does not sound as good as the first. With grid mode enabled and the units of the grid set to measures, you can easily highlight the first chorus, copy it, and then paste it to the top of the second chorus.

Grid mode functions similarly to slip mode, only it restricts where you can place the beginning and end of the cursor. You can set it to measures, which will only let you select one or more complete measures, if you are copying and pasting parts. There are also different rhythm resolutions for grid mode, so select the resolution that best suits the selections you are wishing to edit.

EDITING IN SHUFFLE MODE

Shuffle mode is useful when editing voiceover parts or eliminating sections of a song. Each edit done in shuffle mode will affect the duration of the track. If a selection is cut from the audio, the region to the right will snap to the edit point. If a selection of audio is copied and then pasted using shuffle mode, the audio to the right of the edit point will remain intact; it will just be moved to the right, further down the timeline. In slip mode, the pasting would result in the audio being replaced by the pasted regions.

When editing voiceover tracks, shuffle mode can be useful for removing breaths or false starts from a continuous track. When using shuffle mode to rearrange songs, you have the ability to cut a region across all of the tracks, and then paste it into a different location, without being placed on top of regions. This will help to keep the continuity of the tracks.

Zero-crossing edits

When editing a track, whether you are splicing vocal parts or tightening up instruments, edits are required to be done at a zero-crossing point or an equal point above or below zero. An analog audio signal consists of an alternating current voltage. This mimics the compression and rarefaction of acoustical sound waves. This is represented in the digital domain by having the audio signal's voltage captured as an alternating signal as well.

Any edit that creates a discontinuity in the alternating signal will create an audible pop or click, as the audio signal suddenly has to jump to a new level (Figures 1.15 and 1.16). Following the laws of physics, it is impossible for a signal to jump to a completely different level instantly. It is this instant jump in level that creates the audible pop or click. This is why it is important to do all of your editing in such a way that the different regions meet at a common point. Editing regions is usually done at a zero-crossing point; however, a continuous signal can be edited in such a way that there is no jump in signal level or direction. This can be done at the top or bottom of a waveform, as well as a part of the

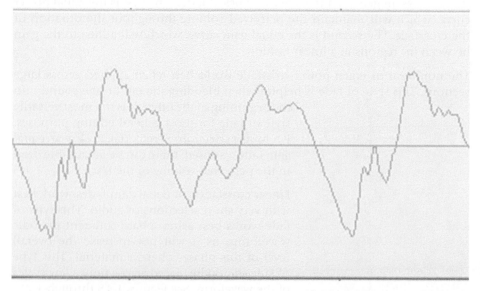

FIGURE 1.15
A recorded waveform
with a continuous signal.

FIGURE 1.16
An abrupt edit of a waveform, creating a discontinuity of the signal, resulting in an audible click.

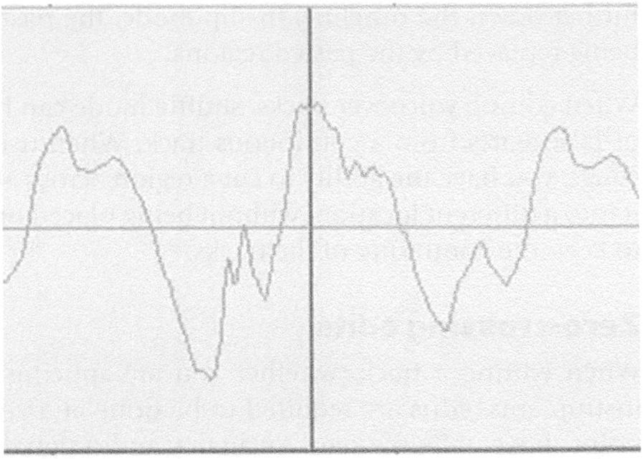

FIGURE 1.17
An edit of a waveform maintaining the amplitude and direction of the waveform.

directional slope. Editing these tracks can be done "peak to peak," aligning the peaks on either side of the zero point for a clean edit (Figure 1.17).

Using crossfades

Crossfading is a means to smooth any editing done between two regions. It will eliminate any noncontinuous edits. There are many different types of fades that can be used between audio regions, and most DAWs will allow you to fine-tune these curves. In general, there are two types of fades. The first is the equal power curve, which will maintain the perceived volume throughout the duration of the crossfade. The second is the equal gain curve, which will adjust to the gain between the regions in a linear fashion.

The nonlinear or equal power crossfade works best when applied across large sections. This style of fade is helpful when blending the tail of one sound into the beginning of the other. It is the most versatile style of fade for most general editing purposes. For longer selections of material, if a linear, equal gain fade was used, there can be an audible drop in the perceived volume of the track.

FIGURE 1.18
A short edit of a phase-coherent waveform, using an equal power curve, resulting in an amplified crossfade.

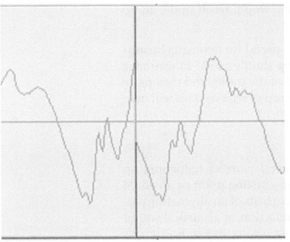

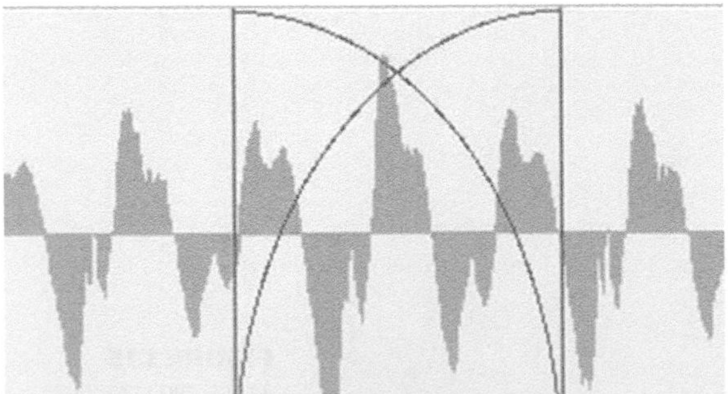

Linear crossfades, or equal gain fades, work best with very short selections of audio. This type of fade works best across phase-coherent periodic waveforms, as it will not increase the overall level of this phase-coherent material. This type of fade should be placed across only a few cycles of the waveform. See Figures 1.18 through 1.21.

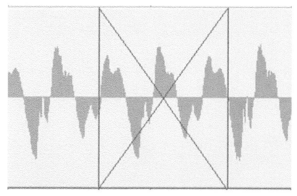

FIGURE 1.19
A short edit of a phase-coherent waveform, using an equal gain curve, resulting in a consistent level through the crossfade.

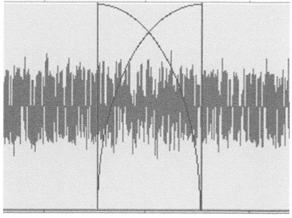

FIGURE 1.20
A long crossfade using an equal power curve, resulting in no major change in level through the crossfade.

A FINAL WORD ON GETTING A GREAT START IN THE STUDIO

FIGURE 1.21
A long crossfade using an equal gain curve, resulting in a noticeable dip in volume.

Getting a great start on a recording is important, as the rest of the recording process will suffer and require extra time and money to fix mistakes that were originally captured during the initial recording. Having to go through and find a way to eliminate bleed from a click track or improve on poor microphone technique can use up time that is better spent on mixing or adding interesting production tracks.

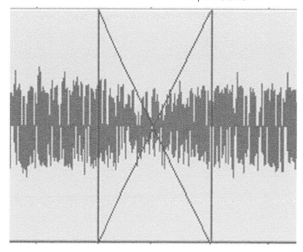

CHAPTER 2
Timing Correction

INTRODUCTION TO TIMING CORRECTION

Correcting the timing of a piece of music, whether it be the drums, percussion, or merely tightening up doubled vocal parts, is one of the most important things that can be done to improve the quality of a recording. If faced with the choice of having a track be out of pitch or out of time in a recording, it is less appealing to the ear to have something out of time. Having parts that are out of time denotes a sloppiness to the performance, which translates to a sloppy recording. The ear is generally more forgiving to pitch, within reason, than timing.

In the world of editing tracks visually, it becomes easy to edit the timing so that tracks "look right." It is important to realize that there is musicality to the slight fluctuations of timing as an artistic license. Some parts sound better slightly behind the beat. When professional musicians are playing together, they know which side of the beat to be on for the style of music and their particular instrument. This is another instance where studying the style of music that you are recording will make you a better engineer.

There are two main types of timing correction. The most common method creates edits in the audio regions and moves them so that the transients are then on the appropriate beats. This can be done either manually or through software features that automate this process, such as Digidesign's Beat Detective.

The other main type of timing correction will actually stretch or compress the time of regions to make them fit the tempo of the song. This can create problems with artifacts created by the digital processing that is used to expand or contract the pitch.

Tools for timing correction

Most timing-correction software tools are proprietary to their host digital audio workstation (DAW) software. Most plug-ins used in this book are available across multiple platforms, as they function as an insert across a track. With timing correction, it is more a feature of the workstation software, as most of the handling of the audio does not take place in real time; they are merely editing features. This chapter will give you an overview of the main features used in this book.

Time compression and expansion

Depending on the algorithms used by a particular time-compression and -expansion feature, there may be artifacts created by the process. This is particularly more noticeable in time expansion as opposed to compression.

Percussive sounds have a specific transient that denotes the beginning of the sound. When these percussive sounds are stretched, oftentimes the transient can be smeared or even replicated to create the lengthened sound. A short time expansion can create a smeared transient. If the transient is smeared, the percussive nature of the sound can be diminished. With even longer time expansion, the transient can be replicated, creating a flam in the sound.

In Figure 2.1, there are five versions of the same audio file, which is a snare drum. It has a solid transient but a moderate decay to the sound. The upper track is the original track. The second version is time expanded by 100 percent, where you can see a double transient created. The third version of the snare drum is time expanded by 50 percent, where you can see a smearing of the transient. Take note

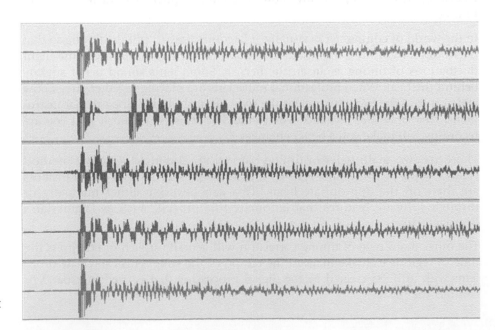

FIGURE 2.1
Five tracks, displaying the effect of time compression and expansion with different algorithms.

that this will create problems in the sound, as it will not cut through the mix as a snare drum should. The fourth version has the snare drum expanded by 100 percent, but with an algorithm that is specifically designed for percussive sounds. This will maintain the transient of the sound but increase the decay. The fifth version of the snare drum is time compressed by 50 percent, and you can see little effect to the transient.

Manual timing correction

When correcting the timing of a performance, there are a couple of different ways that various DAWs can handle the job. The first method that you will usually encounter is manually correcting the timing of a performance. This works well for subtle mistakes, such as the bass player hitting a note slightly earlier. A simple edit and nudging of the audio region will take care of this.

In Figure 2.2, you can see that the second note of the lower track, which is the bass, is behind the beat when compared to the kick drum hit above it. This is a classic example when a simple region nudge edit will correct this issue.

In Figure 2.3, you can see that the region of that note is separated from the surrounding notes, allowing for the region to be nudged to its appropriate position.

The region has been nudged, as seen in Figure 2.4; however, there is a gap between the end of the note and where the next note begins. This is where you will need to make a decision as to how to correct the empty space. In some

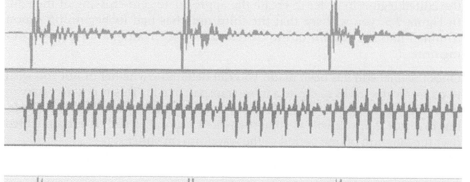

FIGURE 2.2
Kick drum (top), and bass line with the second note out of time when compared to the kick drum (bottom).

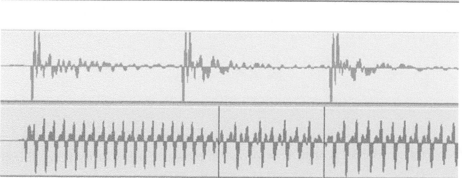

FIGURE 2.3
The region in the bass track has been separated in order to shift the note.

FIGURE 2.4
The region of the bass note has been nudged to be in time with the kick drum.

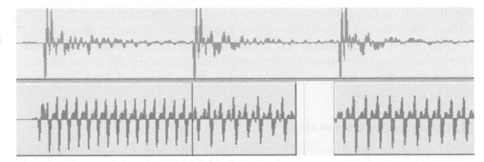

situations, the way that the edit is nudged, without any region adjustment, will sound just fine among all of the other tracks. The best way to make that determination is to listen to all of the other tracks playing at the same time to determine if the gap is noticeable or not.

Sometimes it is best to err on the side of caution and to correct the empty space. This will cover the edit, in the event that the song undergoes some rearranging in the future that would make the empty gap more apparent to the listener.

The first method of correcting this gap is to appropriately drag the end of the note to the right of the edit, back into the edited region. This edit will only work if the edited region has sufficient duration so that the dragged region can overlap during the sustained portion of the region. This is a good place to keep in mind the principles of utilizing zero-crossings for the edit. You may have to slightly nudge the timing of the initial region to create an overlap of the edited region in order to create the appropriate zero-crossing of the edit. In Figure 2.5, you will see that the third note has had its beginning region trimmed into the sustain of the corrected region, creating a longer, sustaining note.

Once the final edit has been made, you can determine whether or not you wish to crossfade the tracks to create a more transparent edit. This is not necessary to do in the beginning of the edited region, since it begins on an attack of the note, but in Figure 2.6, you will see that the edited region has had its sustain edited over the duration of one period of the waveform on either side.

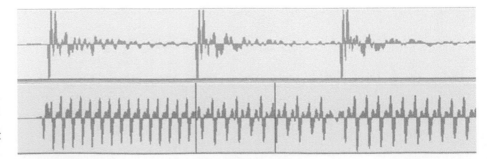

FIGURE 2.5
The corrected bass note has had the end region nudged earlier to correct the gap.

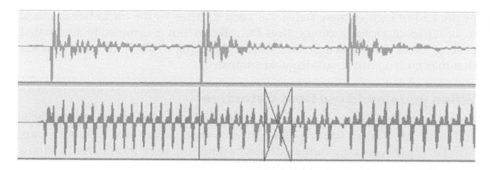

FIGURE 2.6
The nudged end region
has been crossfaded to
create a smoother edit.

BEAT DETECTIVE

One of the most powerful automated editing features added to Digidesign's Pro Tools in recent years is Beat Detective. Originally only available for the HD system, it is now available as a single-track feature in Pro Tools LE, or as the multitrack version found in HD. You can add the multitrack feature to Pro Tools LE with the purchase of the Music Production Toolkit, which also adds a whole host of other features, such as SoundReplacer and the ability to have a session with 48 audio tracks.

Beat Detective is a powerful editing tool that is used to tighten up rhythmic tracks according to the tempo map in Pro Tools. It can also be used to generate a tempo map from an existing performance.

Beat Detective accomplishes this editing by cutting up the regions according to where it detects the transients. It then moves the regions that it has created and places them in the correct time according to the tempo map in Pro Tools.

Steps of Beat Detective

When editing drum tracks with Beat Detective, there are basically three steps, with an optional fourth in the operation of the software to create the appropriate tracks.

- *Region separation*. Region separation will identify the beats performed by the drummer and separate them at the beginning of each selected transient.
- *Region conform*. Region conform then takes these separated regions and moves them to the appropriate bar, beat, or sub-beat in the measures.
- *Edit smoothing*. Edit smoothing can then move the ends of the regions to the adjacent edit so that there will not be any silence between some of the edits. In addition, you have the option of adding a crossfade at each of the edit points.
- *Consolidation of the tracks (optional)*.

At the end of each of these steps, you need to listen to the tracks being edited in order to check for accuracy. Beat Detective is not a completely automated process, but a user-guided process. Careful listening will help to avoid artifacts that may crop up due to misdiagnosed transients.

When you should use Beat Detective

Beat Detective can be used to generate a tempo map, as mentioned in Chapter 1, and it can also be used to put performances in time with an existing tempo map. Beat Detective is great for putting drums in time with the click track or locking a percussion track in time with the drums.

Beat Detective should be used right after basic tracking, before any of the overdubs have been recorded. If the other instruments are following the drum tracks, and you tighten the drum tracks to the tempo map of the song, by correcting the timing of the drums, the other instruments will sound out of time. Keep this in mind when planning your session so that you do not have parts that are out of time, in which case you will need to rerecord over-dubs again.

Saving a new session

When doing any semidestructive editing, it is always good to save a new session file in your session folder. Also, it is a good habit to get into to save a new session file every time you do something major in Pro Tools, from editing to mixing. A helpful method would be to increment a number and add a note as to what is contained in the new session. If you are working on "Song A 1," save a new version as "Song A 2-Beat Detective." This will help you go back at any point in the future in case something gets messed up along the line.

Grouping the drum tracks

Beat Detective will work across multiple drum tracks if you have the multi-track Beat Detective with your software. In order for it to work across all of the drum tracks, they all need to be selected. The easiest way of doing this is to put them all in one group by selecting the names of all of the tracks, then selecting "Group ..." from the Track menu.

Duplicating the playlist

It is always good to create a safety net for yourself when doing semidestructive editing to your tracks. This can be accomplished by duplicating the playlist in Pro Tools so that you can always go back to an unedited version. It will be helpful to duplicate the playlist at the beginning and before starting each phase of using Beat Detective. This way, if you make a mistake, it is easy to copy and paste a previously unedited version on top of your active playlist.

To duplicate the playlist, make sure that the tracks you are using in Beat Detective are in a single group. Click on the triangle to the right of one of the track names and select "Duplicate Playlist." This will create a duplicate of the playlist so that you can now always go back to the original. The name of the playlist increases the name by .01, so keep track of what playlist you are working on. See Figure 2.7 for an example.

Creating sections to edit

In much the same way that we created a tempo map from the drummer's performance in Chapter 1, using four or eight bars at a time, we will use the same smaller sections to correct the timing of the drums here.

FIGURE 2.7
Duplicating the playlist in Pro Tools.

The reason to choose shorter selections is that Beat Detective gives you a wealth of options for resolution. You can adjust the resolution from a single bar down to 32nd notes. Since not every measure that the drummer plays will contain the same rhythmic resolution, it is best to break up these sections into the exact rhythm performed in the bar. A simple drum groove may only contain quarter notes or 8th notes, but when a drum fill is played, there can be 16th notes in the bars that contain the fills. Beat Detective will work best when given the most accurate rhythmic information according to the content of the track.

In the example in this book, the tempo of the session is 68 beats per minute. With a slower tempo, inexperienced drummers have a tendency to rush the beat, so the use of Beat Detective will be important in creating a solid and steady feel to the song. Because of the slow tempo, we will be selecting two measures to edit at a time.

Selecting sections works well by separating the two-bar region precisely. This is done by zooming in and placing the cursor right before the first hit of the kick drum as it hits right on the downbeat.

Selecting the region for editing

When you run the "Region Separation" phase of Beat Detective, it will automatically separate the region for you; however, it can be helpful to manually select the beginning and end points of the measures you are working with in small chunks to keep track of the small sections as you work through the session.

In Figure 2.8, you can see that the region has been manually separated at the beginning of bar 70, where the drums begin. The region can be manually separated by pressing "Command-E" (Mac) or "Control-E" (PC). If single-key shortcuts are enabled, simply pressing "B" will create the appropriate edit.

FIGURE 2.8

The drum tracks' regions have been separated precisely before the first transient.

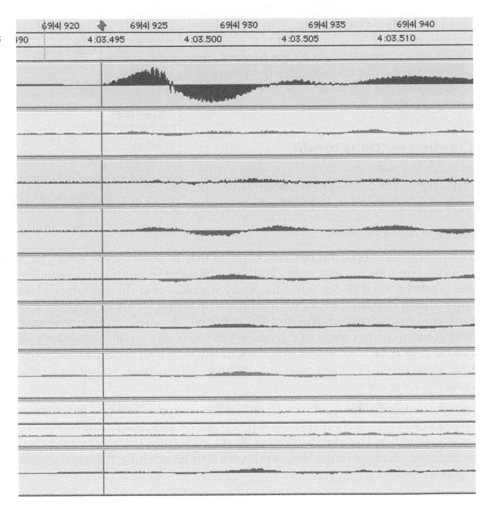

The beginning of bar 72 is then separated in the same way so that we have the exact two bars selected that the drummer played. Now we want that section to conform in time to the same corresponding four bars in Pro Tools.

In Figure 2.9, you will see the region for the two bars, as performed by the drummer, separated but not edited yet. Make sure those regions are selected by double-clicking in the middle of the separated region. From this section, we can either manually enter in bars 70 and 72 for the Start Bar and End Bar, or we can take the highlighted tracks and click on "Capture Selection."

To begin the editing process, we will need to make sure that "Region Separation" is selected under "Operation" on the left (Figure 2.10). Upon listening to the track, determine the shortest rhythm that is contained in the performance. Beat Detective will be more accurate with the longest timing selection that is used in the performance.

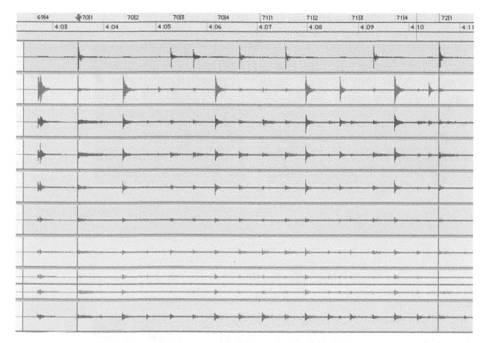

FIGURE 2.9
Two measures of the drum tracks have been separated.

Tabbing to transients

Pro Tools has a "tab to transients" feature that you may be tempted to use, as it will advance the cursor to the next transient that it detects. Even though this is given as an option, it is more accurate to do this manually, as "tab to transients" can place the cursor in the middle of the transient, and you will usually want to place your edit point right before the transient.

FIGURE 2.10
Selecting the Region Separation operation in Beat Detective.

Selecting the appropriate resolution for selection, analysis, and separation

After you have the appropriate measures separated and selected, it is now time to get Beat Detective to automatically separate the regions (Figure 2.11).

In this selection you can see that the two bars contain 16th notes, so we will select "Contains" and set it to 16th notes. Any setting shorter than that can identify mistimed 16th notes as being 32nd notes. Additionally, if the selection contains 16th notes, and you have selected that it only contains 8th notes, then Beat Detective can possibly identify 16th note transients and place them on an 8th note beat when it comes time to conform.

Operation: Audio

○ Bar | Beat Marker Generation

○ Groove Template Extraction

◉ Region Separation

○ Region Conform

○ Edit Smoothing

BEAT DETECTIVE

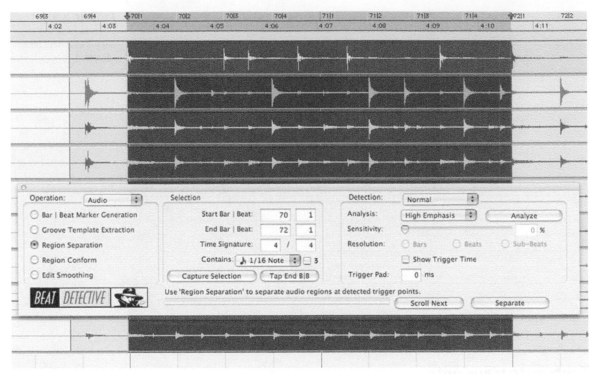

FIGURE 2.11
The two-measure region, ready to be analyzed and separated.

Choosing the detection method

On the right side of the Beat Detective window, you will see three potential options under the "Analysis" selection: "High Emphasis," "Low Emphasis," and "Enhanced Resolution." Selecting either high or low emphasis will focus on either high or low frequencies, respectively. "Enhanced Resolution," found only in Pro Tools 7.4 and higher, has a more complex method of analysis and should be used as your default setting when editing drum tracks with Beat Detective.

Analyzing the audio

Once you have the settings and selection that you need when using Beat Detective, the software will need to analyze the audio, and then you can adjust the sensitivity and resolution accordingly. This is done by pressing the "Analyze" button, and the previously grayed-out sensitivity bar and resolution buttons will become active.

Adjusting the sensitivity for the material

You have the choice in adjusting the sensitivity of Beat Detective by selecting the appropriate option under the detection menu. The sloppier the performance, the higher you are going to want your resolution to be. There are three choices for setting the resolution of Beat Detective.

1. The first is using "Bars," which will place a thick indicator, called a beat trigger, over the selected audio-denoting bars. If you are looking to just tighten up the downbeat of each measure, without affecting the beats in the middle of the bar, then select "Bars."
2. The second choice is "Beats," which will allow Beat Detective to add quarter notes, in addition to bars. These are denoted by a thinner line than the bar triggers, which makes it easier to visualize where Beat Detective will separate the regions.
3. The third choice for resolution is "Sub-Beats," which will then include 8th notes through 16th notes, depending on the selection you have chosen from the "Contains" menu.

After you have selected the appropriate resolution for your selection, you can bring the sensitivity slider up until you see all of the beats in the performance selected with a beat trigger. If you are not seeing all of the beats indicated by Beat Detective, then you may need to choose a smaller selection.

If you have the "Show Trigger Time" button selected, you will see what beats that Beat Detective is identifying the transients to be. These numbers are in the form of Bars | Beats | Sub-Beats, with the Sub-Beats being shown as Pro Tools' MIDI resolution. This MIDI resolution indicates what type of sub-beat is being identified on a scale of 960 ticks per quarter note. If Beat Detective is showing 480 ticks, then it is indicating that it identifies that sub-beat as being an 8th note. A display of 720 ticks would indicate that Beat Detective is identifying that beat as being the fourth 16th note of that particular beat.

Accurately selecting the sensitivity

In general, the lowest that you can bring the sensitivity slider up to where you can see all of the bars, the more accurate it will be. There may be a point when you keep raising the sensitivity, and you will see all of the bars and beats jump to an inaccurate portion of the performance. Increasing the sensitivity will detect softer transients such as hi-hat hits.

In Figure 2.12, you can see that with a sensitivity setting of 19 percent, all of the beats, bars, and sub-beats have been identified. If you look at Figure 2.13, you will see that raising the sensitivity higher, up to 71 percent, indicates that it is inaccurately picking up a 16th note in the decay after the downbeat of the second bar. This can create an audible glitch to the sound after everything has been snapped to its new accurate location.

At this point, it is good to play back the selection to make sure that the beats that you are hearing in the drum parts are indicated by Beat Detective before separating the regions. If everything sounds correct, then press "Separate," and Pro Tools will put edits in all of the regions that it has indicated, and it will adjust them to the correct timing.

Adjusting the beat triggers

The beat triggers that Beat Detective places in the selected region, once the sensitivity has been raised, can be edited. If there are false triggers placed in the selection, those can be deleted by using the Grabber tool and Alt-clicking (PC)

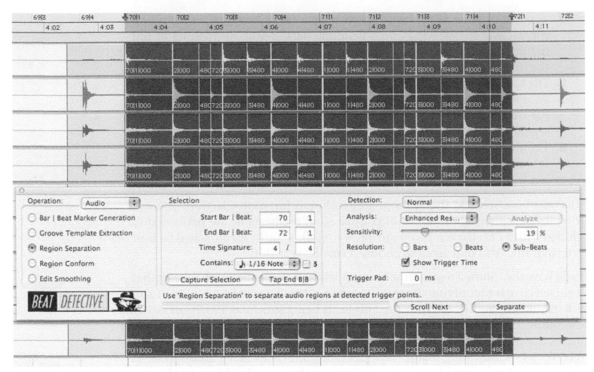

FIGURE 2.12
The region separation markers accurately detecting the beats.

or Option-clicking (Mac) them. In turn, if the transients are not detected as accurately as you may want, you can move the beat triggers by using the Grabber tool and dragging them left or right to the beginning of the transient.

Conforming the regions

After Beat Detective has placed edit points on all of the beats that you are wishing to correct, it is then time to move on to the Region Conform operation by clicking on that button on the left screen (Figure 2.14). Here we have a new set of options on the right side of Beat Detective's window. With "Conform" set to "Standard," which is the correct setting for timing correction, you will see three sliders.

DETERMINING HOW MUCH TO CONFORM

If you want precise accuracy to the drum parts, check the "Strength" box and slide it all the way up to 100 percent, leaving the "Exclude Within" and "Swing" boxes unchecked (Figure 2.15). This will move the separated regions precisely to the tempo map. If you want to maintain some of the tempo fluctuations, you can lower the strength as you see fit.

MAINTAINING SOME OF THE PERFORMANCE

It should be noted that, if you use a setting lower than 100 percent, any sequencing will follow the tempo map and not the drummer's performance. If you want

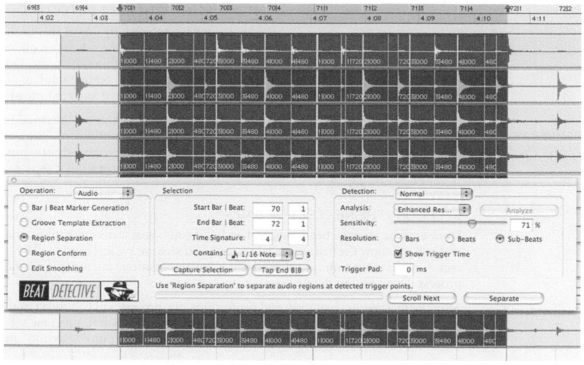

FIGURE 2.13
The region markers have a few noticeable inaccurate detections, due to an increased sensitivity.

to maintain some of the imperfections and still have loops follow the drummer's performance, you will need to adjust the tempo map according to the performance after all of the timing corrections have been made. This is explained in detail in Chapter 1.

THE FINAL CONFORMATION OF THE REGIONS

When you are conforming the selections, the edited selections are moved to their new precisely timed location. In Figure 2.16, you will see that the regions have been conformed. This is a good time to listen to the drum parts with the click track added to the mix so that you can hear if everything has been conformed accurately. You will hear the sounds cut out in the spaces where Beat Detective has created gaps by shuffling the regions around. Do not worry about this, as this will be cleared up in the "Region Conform" stage of using Beat Detective.

FIGURE 2.14
Selecting the Region Conform mode of operation.

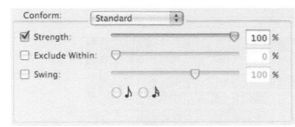

FIGURE 2.15
The conforming strength set to 100 percent for maximum accuracy.

CONTINUING ON WITH BEAT DETECTIVE

Now that you have successfully edited the drum parts for a small section of the track, move on to the next small section throughout the piece. Do not use the "Edit Smoothing" feature until all of the regions have been separated and conformed.

After you have edited this first section, the remaining sections are much easier, as Beat Detective keeps the settings of "Region Separation" and "Region Conform." You may need to make slight adjustments to the sensitivity slider and resolution in a few places, but most similar sections should be easily edited.

SELECTING THE NEXT SET OF BARS

Since we separated the beginning and end of the first Beat Detective regions, the end region is now the first region edit of our next section. We just need to place a region separation at our next chosen end point and continue on, using the same process that we did in the first section.

After the first region separation and conforming, the process will go much faster. After selecting the end point for the next section, double-click in this new section, so that it is all highlighted in Pro Tools, and click "Capture Selection" in

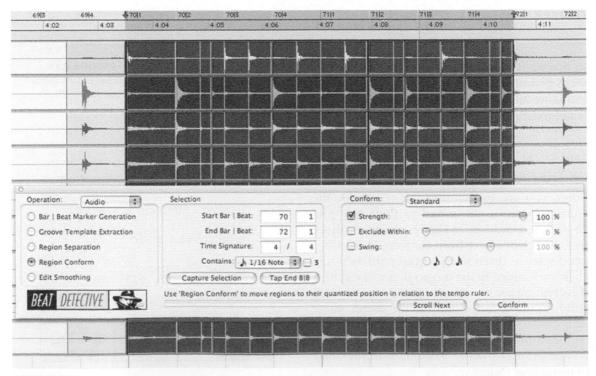

FIGURE 2.16
The separated regions have been conformed at 100 percent strength, with gaps between the adjusted regions.

Beat Detective. This should have adjusted the Start Bar | Beat and End Bar | Beat appropriately, but double check to make sure.

Continue on with the separation and conforming as you did before until you reach the end of the drum tracks that you are using Beat Detective on.

A NOTE ABOUT CONFORMING

Sometimes when conforming regions with Beat Detective, it will shift the last region to the right, which will cover up the transient of the downbeat of the next bar. This happens when the drummer is ahead of the beat. This is easily rectified by using the Trim tool on the unedited track and trimming the unedited region on the right to the left until you see the transient of the downbeat for the next bar.

As you can see in Figure 2.17, the transient downbeat is covered up by the conformed region at the beginning of bar 74. In Figure 2.18, the transient on the right has been brought back by trimming the region on the right over until it has reappeared.

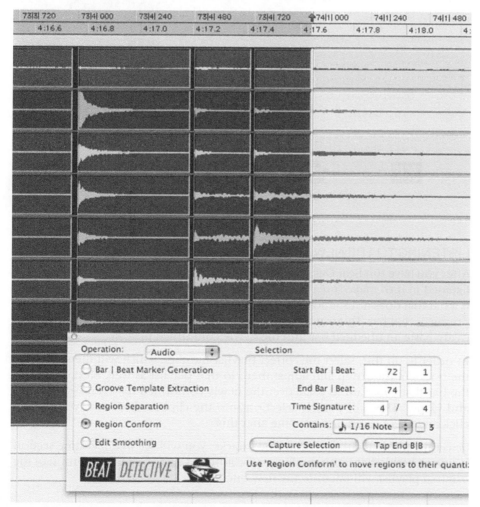

FIGURE 2.17
The transient for the downbeat of bar 74 has been covered up by the conformed regions.

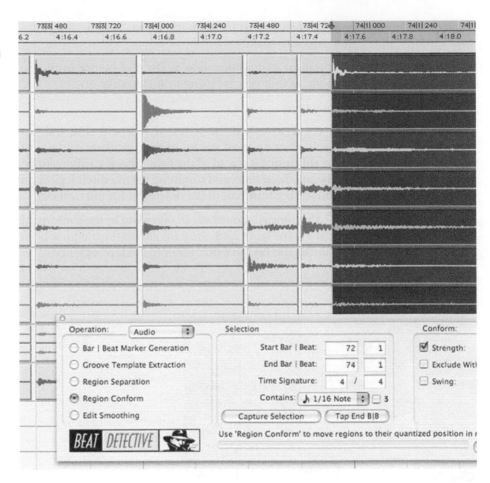

FIGURE 2.18

The transient for bar 74 has been revealed using the Trim tool to drag the start earlier.

LISTENING TO THE NEWLY CONFORMED TRACK

After you have run Beat Detective across the entire track, make sure that you have listened to it to make sure that everything has been adjusted accurately. If there is a bar that does not sound quite right, you may need to individually correct the timing of that bar. This is where the duplicated playlist comes in handy. You can easily paste an unedited bar into the newly conformed playlist.

In order to do this, select the correct playlist number, which will then select all the tracks in the group and adjust them to that playlist as well. Next, highlight the bar that you want to paste over the newly conformed track, and go to "Edit and Copy." This will put the selection into the clipboard. Be sure you do not click anywhere else in the timeline after this.

Select the playlist with the conformed tracks. You will see that the same section has been highlighted. You can just go to "Edit and Paste," and that portion of the original playlist will fall onto the new one.

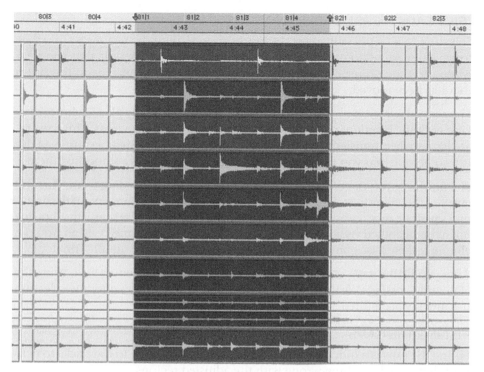

FIGURE 2.19
A previous unedited
playlist has been pasted
on top of a section
previously edited with
Beat Detective.

In Figure 2.19, you can see the unedited version pasted into the conformed version. This gives you the ability to always go back to an unedited version if you ever change your mind in the future about having used Beat Detective to do the edits.

Edit smoothing

Before you move on to any edit smoothing, this is another good time to duplicate the playlist of the conformed tracks, so now you have three versions of the drum tracks: an unedited version, a conformed version, and a smoothed version.

Edit smoothing consists of only two choices: "Fill Gaps" and "Fill and Crossfade." When selecting "Fill and Crossfade," you will also have the option to adjust the crossfade length in milliseconds.

FIGURE 2.20
Selecting "Fill and
Crossfade" for the edit
smoothing operation.

SELECTING THE ENTIRE REGION

Make sure that you have not put any crossfades in the middle of the conformed tracks. Beat Detective will not perform its smoothing operation if there are any crossfades, because the smoothing operation will adjust the boundaries of the regions, and crossfades have no additional boundaries to adjust.

Smoothing

○ Fill Gaps

◉ Fill And Crossfade

Crossfade Length: [1] ms

"FILL GAPS" VERSUS "FILL AND CROSSFADE"

Selecting "Fill Gaps" will adjust the boundaries of all of the regions so that they touch the next adjacent region. This will eliminate the dropouts that you hear when there are no regions in between some of the transients, due to the audio that has been moved.

The second selection of "Fill and Crossfade" will additionally place a short cross-fade with the length of your choosing at each of these region edits. This can create issues when transients are crossfaded into another transient, creating a false hit, so be sure that you listen to the track right after you have smoothed them.

The first thing you can do is to try and fill gaps without crossfading them. This will give you the ability to make sure that there are no false double transients. If the drummer's performance is way off, there is the possibility of these false transients showing up in the smoothed version that you did not hear in the conformed version. You can eliminate these false transients by manually adjusting the edit with the Trim tool.

After you fill the gaps, you can optionally try to crossfade them as well, using the "Fill and Crossfade" feature, as it will only add the crossfades after the selection has already been filled.

FILLING THE GAPS

Filling the gaps is a quick process. Make sure that you listen to the tracks to confirm that there is nothing wrong with the adjusted regions.

In Figure 2.21, you can see the drum tracks that have not been smoothed. In Figure 2.22, you will see that the gaps between the regions have been smoothed by having their boundaries adjusted.

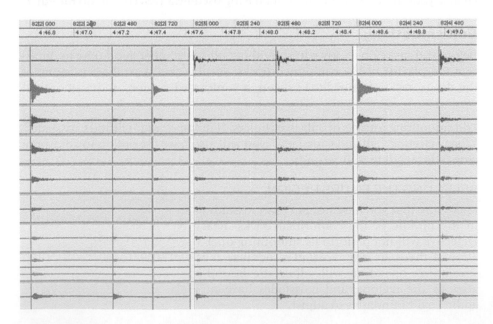

FIGURE 2.21
Regions separated and conformed, but not yet smoothed.

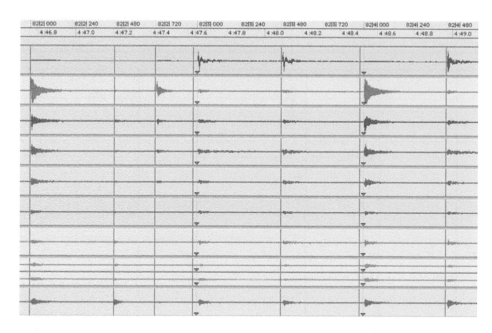

FIGURE 2.22
The conformed regions have been smoothed, eliminating any gaps between regions.

CROSSFADES

Adding a crossfade to the selection actually writes a new audio file that consists of that crossfaded region. This saves processing by eliminating the need for the software to perform the crossfading in real time and takes up only a little bit of hard disk space. Crossfading can eliminate any pops and clicks that you may hear due to non-zero-crossing edits of the audio.

Choosing to use the crossfade feature requires listening to the sound to see if it smoothes the sound out better, or if it is creating more artifacts, with the added bonus of making the sound smoother. If you are using the crossfade, you have the option of pasting in an unsmoothed version from the previously used playlist in order to eliminate any artifacts.

Choosing a very small crossfade length such as 1 ms will have less of a chance to blend in a false transient, but it will eliminate any pops and clicks that may arise from conforming the regions.

In Figure 2.23 of the same selection, the crossfade feature was added to the same previous selection. You can see the very small crossfades that are over each of the edit points.

Consolidating the edited tracks

Once you have run all the phases of Beat Detective and there are no artifacts to the audio, you can choose to consolidate the selection. Before you do this, be sure that you duplicate the playlist so that you can always go back to this version in case you missed an added false transient, or the track is not smoothed the way that you want.

FIGURE 2.23
The conformed regions with a 1-ms crossfade added in between.

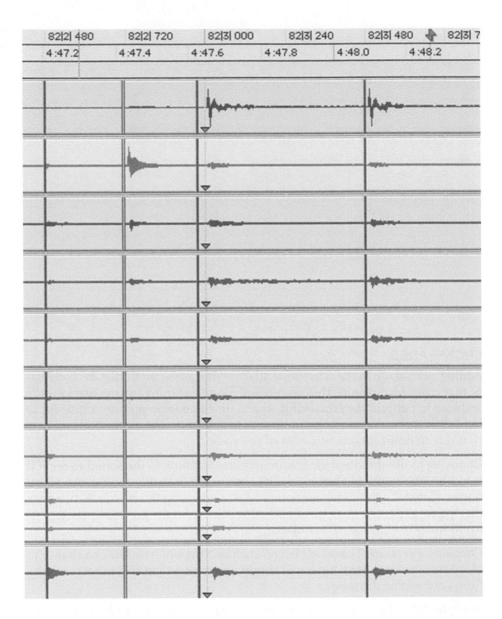

Consolidating the tracks rewrites all the selected audio into a continuous audio file. Since Beat Detective generates hundreds of edits and crossfades, there are many files that Pro Tools is trying to read in real time as you play through the track. This will take up processing as well as bandwidth reading from your hard drive. Consolidation is the solution to this issue.

In order to consolidate the tracks, highlight all the contents from beginning to end. Go to the Edit menu and select "Consolidate," and Pro Tools will begin consolidating all of the highlighted audio regions. Since this is writing new audio tracks, it will take up more hard disk space and will add additional files to your Audio Files window.

Speeding up the beat with Beat Detective

One of the more interesting tricks that you can do with Beat Detective is to use it to adjust the speed of a performance.

An easy thing to do is to take a two-measure selection and convince Beat Detective that it is a one-measure selection. In this example, we will use the first two measures that we originally applied Beat Detective to. When going to the Region Separation operation, tell Beat Detective that the End Bar | Beat is bar 71, even though you have through bar 72 highlighted.

In Figure 2.24, you will see that the "Contains" has been adjusted so that it halves the length of the previous resolution from 16th notes to 32nd notes. This is because we are going to trick Beat Detective into thinking that these two bars are actually one bar.

Go through the analysis and separation like you did before. You may need to increase the sensitivity and do some manual editing of the beat triggers if you are going to try and capture the softer hi-hat notes. Then conform the selection as you have done previously.

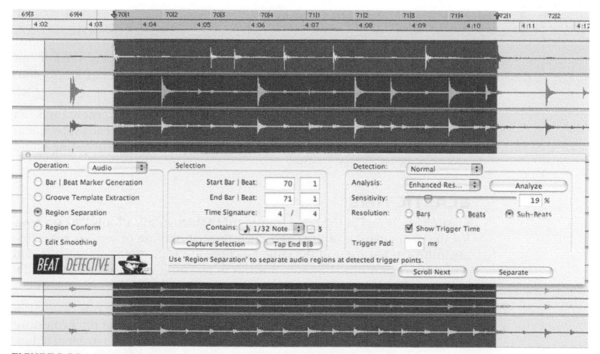

FIGURE 2.24
Forcing Beat Detective to recognize a two-measure selection as one measure.

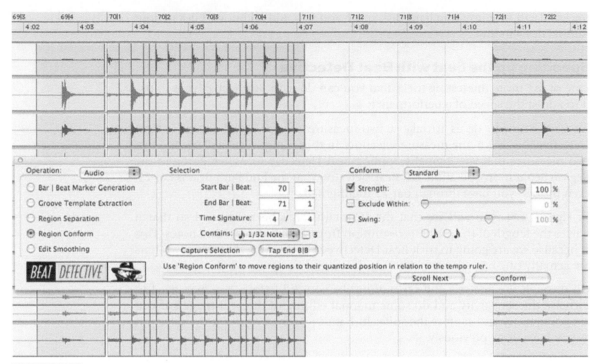

FIGURE 2.25
The two-measure selection has been separated and conformed onto one measure.

In Figure 2.25, you will see that now we have the time-compressed version, but rather than using a time-compression algorithm, Beat Detective has simply doubled the performance through the use of edits to the regions.

This can be an interesting way to accomplish time compression. The sounds can come across as being gated, as the decays are being cut off with each subsequent edit in an unnatural way; however, this can be one of the more interesting sound design methods in your toolbox.

ELASTIC TIME

Elastic Time is one of the main features of Digidesign's Pro Tools 7.4. It allows for the manipulation of time and pitch through the use of various selectable algorithms. It has some of the same functionality as the time-compression and -expansion algorithms; however, it works in real time, which allows for nondestructive adjustments to be made in the audio tracks.

The two most powerful editing functions of Elastic Time that can be used in record production are the alignment of timing of vocals or other pitched instruments, and the correction of the drum and percussion tracks. It may seem redundant to use Elastic Time to correct the drums when you have the use of Beat Detective, but it merely becomes a different tool at your disposal.

Beat Detective works differently than Elastic Time. Beat Detective automatically splices and aligns audio based on the transients detected. Elastic Time is able to stretch and compress the audio based on detected transients. This has the advantage of allowing you to not worry about crossfades and extra transients being created through the smoothing process in Beat Detective.

Elastic Time has different means of analyzing audio. In order to make the most out of Elastic Time, selecting the appropriate algorithm is important. Since Elastic Time functions in real time, it uses up processing for each of the tracks on which it is applied. In essence, it is an optional real-time plug-in with the option of rendering the performance as audio files on your hard drive so that you are not taxing your computer processing unit (CPU).

FIGURE 2.26
Selecting the different algorithms for Elastic Time from the Edit window.

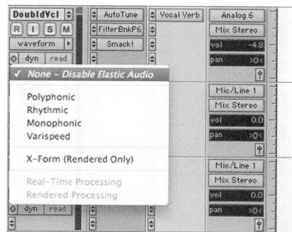

There are three main algorithms in Elastic Time that are used to correct and adjust timing: polyphonic, rhythmic, and monophonic. There is also a varispeed algorithm that can create different effects on the audio by speeding up and slowing down the audio based on the tempo changes of the tempo map in relation to the warp markers.

Engaging Elastic Time is as simple as clicking a selector tab underneath the track's name at the bottom of each track in the edit window. In order to best use Elastic Time, you need to select the algorithm that is best suited for your material (Figure 2.26).

Elastic Time algorithms
POLYPHONIC MODE

The polyphonic algorithm is best suited for tracks that have more than one note playing simultaneously. This algorithm is useful on guitar or keyboard tracks.

RHYTHMIC MODE

The rhythmic algorithm is best suited for drums and percussion. It will keep the transient attacks intact while still being able to manipulate and stretch the audio. This algorithm would not be the best choice for any specifically pitched audio.

MONOPHONIC MODE

The monophonic algorithm will work on a track that has only one note playing at a time. This is best suited for lead and background vocals or individual horn parts.

VARISPEED MODE

The varispeed algorithm will create special effects to the audio. When the audio is being stretched, Elastic Time will slow down the audio track as if it were an analog tape machine. When the audio is being compressed, the sound will be pitched up.

Analyzing the audio track

Elastic Time, once engaged by selecting an algorithm, will then analyze the audio on the track. If there are multiple tracks in a group it may take some time for Elastic Time to complete its analysis. The track will be grayed out until the analysis is complete.

This analysis does not analyze the frequency content of the audio track, rather it merely analyzes the track for transients. This is used if you are specifically going to quantize the track to the tempo map.

WARP VIEW

Once an algorithm has been selected and the audio track has been analyzed, you now have the option of switching to view the track in warp and analysis view (Figure 2.27). Looking at the track in warp view, you now have the ability to insert warp markers, and can now go about stretching and compressing the audio.

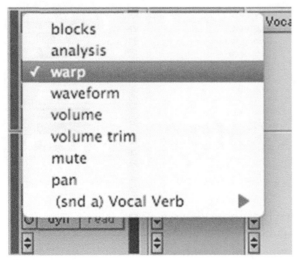

FIGURE 2.27
Adjusting the view of the track to create and edit warp markers.

Aligning vocals using Elastic Time

Anytime you are recording with multiple vocal tracks, there may be a time in which you will need to align either background vocals or a doubled vocal part to the lead vocal. Elastic Time makes a great tool for manually adjusting the timing of these vocal parts to the lead vocal.

In order to align vocal parts using Elastic Time you will need to insert warp markers. If you are editing a track in the warp view, you can either select the Pencil tool and add them in manually or use the Grabber tool and Start-click (PC) or Control-click (Mac). It is easiest to use the Grabber tool, as this is the same tool that is used to drag the warp markers in warp view. Additionally, warp markers can be deleted by simply pressing the Option key and clicking on the warp marker with either the Pencil tool or the Grabber tool.

Elastic Time will stretch and compress any audio on either side of the warp markers. A good habit to get into would be to place a warp marker at the beginning and end of the selection that you are adjusting. This will prevent Elastic Time from compressing or expanding the entire audio track (Figure 2.28).

When using Elastic Time for vocal manipulation, it is best to go and edit the vocals phrase by phrase. Placing a warp marker at the beginning and end of each phrase or note will allow for just the time compression and expansion of those individual phrases (Figure 2.29).

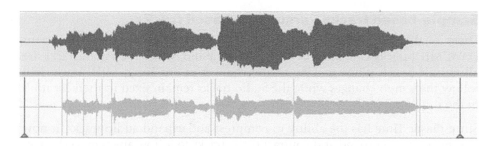

FIGURE 2.28
Inserting warp markers before and after a selection to prevent accidental adjustment of regions to the left and right.

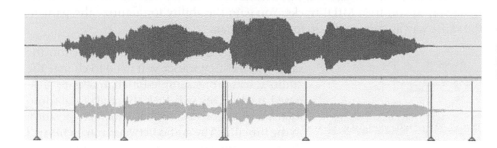

FIGURE 2.29
The warp markers placed at the beginning and ending of each phrase.

After the markers have been placed, select the Grabber tool. From here, it is merely a matter of dragging the markers in such a way that they will line up the beginning and ending of each of the notes or phrases. You may find that you may need to add an extra warp marker in a few locations to get the timing to be exactly what you want. See Figure 2.30.

Using Elastic Time to quantize drum parts

Elastic Time has the ability to effectively quantize drum parts in much the same way that you would quantize MIDI tracks. Pro Tools will analyze the drum parts for transients and then align the transients to the grid based on the resolution that you choose. This works only if the tempo map is accurate to the recorded tracks. If you recorded to a click track from inside Pro Tools, then the tempo map will match up with your editing.

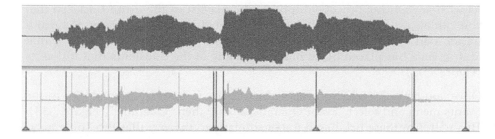

FIGURE 2.30
The warp markers have been adjusted to make the doubled vocal match the original.

Sample-based tracks versus tick-based tracks

There is a difference between working with MIDI tracks and audio tracks in your DAW. MIDI tracks contain only data and no audio, while the audio tracks are all sample based. When making tempo changes in Pro Tools, MIDI tracks will follow the tempo changes while the audio tracks remain fixed in their location in the timeline.

Since Elastic Time has the ability to compress and expand audio tracks, it now has the ability to expand and contract those tracks to follow any tempo changes that you may make. In essence, this means that you can treat audio tracks in the same way as you treat MIDI tracks. You gain the ability to quantize the parts in the same way as you would quantize MIDI data.

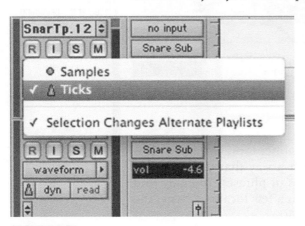

In order to get the audio tracks to function the same way as MIDI tracks, you need to change the audio tracks from sample-based tracks to tick-based tracks. This will lock the audio data into their bar locations as opposed to their position in the timeline. The audio between transients will then be compressed or expanded to adjust to the tempo changes. To switch a track from sample based to tick based, click on the box below the view in the Edit window of the track and change your selection appropriately (Figure 2.31). With all the drum tracks in a group, you can switch them all at once.

FIGURE 2.31
Selecting the track to go from sample based to tick based.

Engaging Elastic Time

To engage Elastic Time, all you need to do is select the rhythmic algorithm in order for it to begin its analysis. Once all the tracks have been analyzed for their transients, the initial quantization of the parts becomes easy. Throughout the course of correcting the time in the drum tracks, you will find it helpful to turn on a click track so that you can make sure that any quantization done is accurate.

QUANTIZING THE DRUM TRACKS AS EVENTS

As with any editing process, duplicate the playlist of the drum tracks so that you can always go back to the starting point. To begin quantizing the drum tracks, start by switching the view to "Analysis." From here you will see that Elastic Time has placed black markers on any of the transients that it detects (Figure 2.32).

Once you have the selection highlighted that you want to quantize, go to the Events menu and select "Event Operations" and then "Quantize" (Figure 2.33). This will treat these tick-based tracks similarly to the way that Pro Tools treats MIDI tracks.

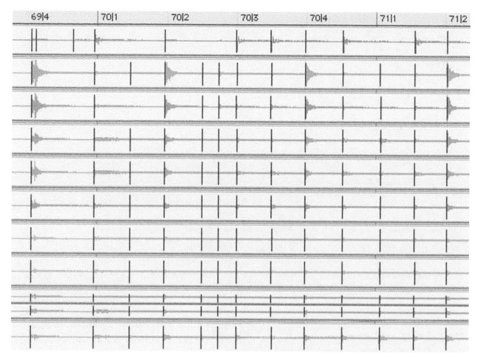

FIGURE 2.32
Transients detected by Elastic Time as black lines.

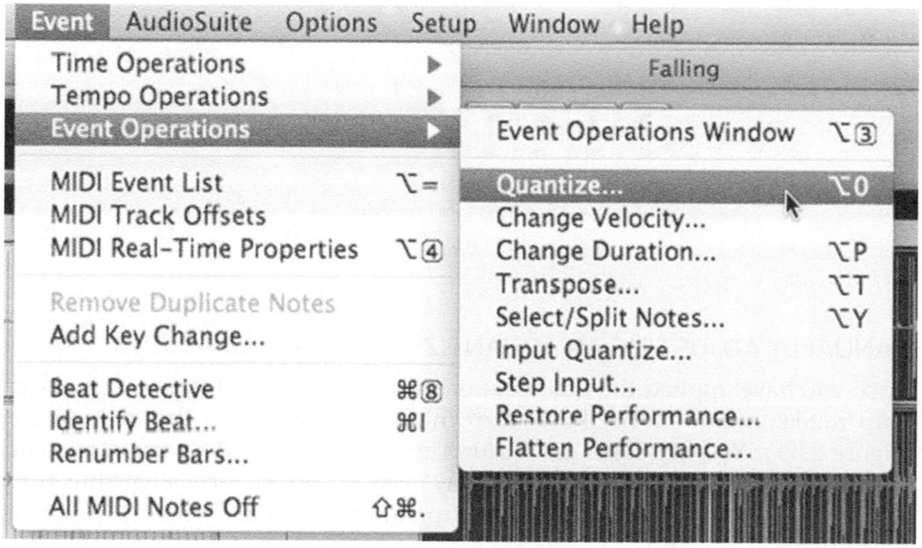

FIGURE 2.33
Selecting "Quantize" from the Events menu.

The "Event Operations" window will then give you the same options that it does if they were MIDI tracks. Select the quantized grid based on the drummer's performance and then the strength of the quantization. You have the option of not quantizing it 100 percent to the tempo map; however, if you want a very precise performance, increase the strength as high as you feel comfortable (Figure 2.34).

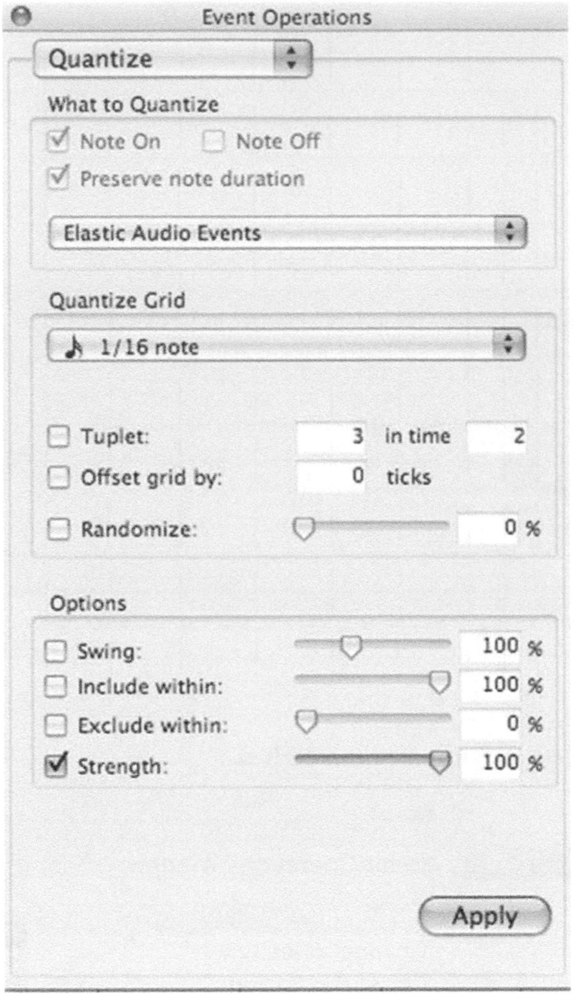

FIGURE 2.34
Quantize settings for
Elastic Time to snap
the audio to the nearest
16th note.

MANUALLY ADJUSTING ANY QUANTIZING MISCALCULATIONS

Once you have applied the quantization across the tracks, Elastic Time places warp markers and adjusts those warp markers according to the tempo map (Figure 2.35). You will find that Elastic Time may not be 100 percent accurate over the entire performance, and you may need to correct a few single hits here or there. Switch to view the tracks in warp, and you can now make the annual adjustments as necessary.

Making corrections to the quantization using Elastic Time is very easy to do. Correcting these errors is merely a matter of manually removing and placing a new warp marker (Figure 2.36). Warp markers can easily be deleted by selecting the Grabber tool and Option-clicking on the misplaced warp marker.

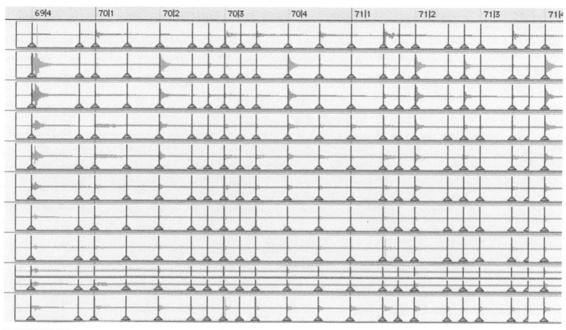

FIGURE 2.35
The warp markers have been moved after quantizing the tracks.

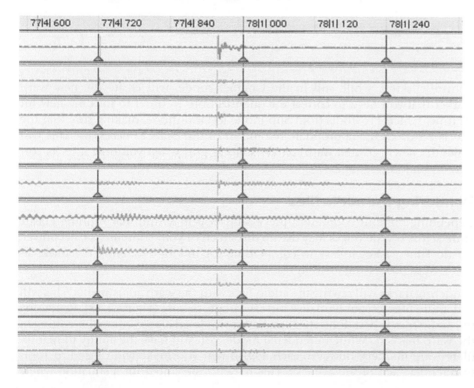

FIGURE 2.36
Warp markers that have missed the transient and need manual adjustment.

FIGURE 2.37
Warp markers that have been adjusted to be on the kick drum hit, and then manually snapped to the correct beat.

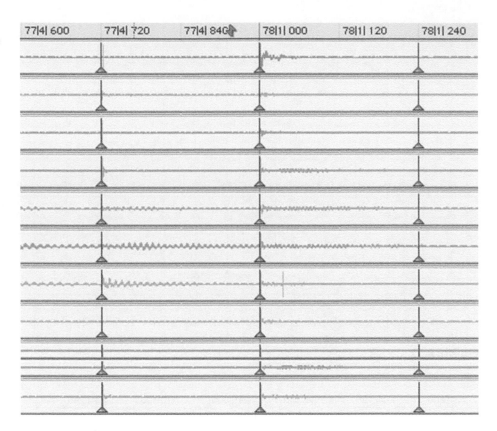

Control-clicking, while still in the Grabber mode, will create the new warp marker that you need to correct the quantization. In order to lock the new hit on time, switch Pro Tools from Slip mode to Grid mode, making sure that the grid is set up to the appropriate resolution of the beats as necessary. This will make the newly created warp marker snap to grid when you drag it to its corrected location (Figure 2.37).

Rendering the audio files

Depending on your computer's processing, you may find that running Elastic Time across several tracks may use up too much of your computer's CPU power. Elastic Time makes it easy to compensate for this by rendering the tracks as opposed to processing them in real time.

Rendering tracks will rewrite the audio files with the warp adjustments. This will add audio files to your hard disk but save in processing (Figure 2.38). Rendering the audio files is best done after you have completed all of your major editing of the warp markers. While in rendered mode, you still have the ability to manipulate warp markers; however, each adjustment you make will cause Pro Tools to rerender each of those tracks. This may not be suited for a situation where you are trying to work quickly, but it is not a big deal if you are only going to need to make a few adjustments to the warp markers.

After you have done several corrections using Elastic Time, if you turn off Elastic Time from the tracks, Pro Tools gives you the option of committing those warp markers to the audio tracks. Unless you are completely done with the project and are archiving the recording, it would be best to keep the tracks in rendered mode so you can still make adjustments easily up until the very end.

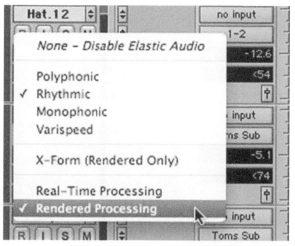

FIGURE 2.38
Selecting "Rendered Processing" to write the audio file and save the computer's CPU.

A FINAL WORD ABOUT TIMING CORRECTION

Correcting any timing issues is important to do before moving on to other production techniques. Since everything will be based on the rhythm of the basic tracks as well as the overdubs, imperfections and timing will have a cumulative effect on the recording. If the overdubbing is taking place on top of mistimed drums or other instruments, it becomes more difficult to recognize whether these overdubbed parts are in time with the track.

A FINAL WORD ABOUT TIMING CORRECTION

CHAPTER 3

Replacing Sounds

REPLACING SOUNDS OVERVIEW

There can be many reasons why you would choose to trigger or replace drum sounds in a recording; for instance, to

- Create a consistency of the kick and snare drums sounds.
- Eliminate bleed from the other instruments.
- Vary the sounds on different songs.

With a sloppy drummer, the different drum hits may be inconsistent throughout the song. Some of this can be minimized through the use of compression, but if you are looking to create a pop record, consistency has been the standard in recent decades.

If you have several microphones on a drum kit, there will be bleed in each of these microphones from the other drums in the room. It can be difficult to get some of the bleed out, even with careful microphone choice and placement. Sometimes the drummer's setup makes things even more difficult, if he or she places the cymbals too close to the drums.

There are stories that abound of producers recording each drum track individually. This can be an extremely time-consuming process, as well as being very difficult for the drummer to accomplish. Sometimes cymbals are recorded separately from the rest of the drums, but again, you are at the mercy of the drummer's ability as well as the studio budget.

Suppose you are working with a drummer and he or she has only a single snare drum. If you want to change the sound of the drums in different songs, the triggering of a different snare drum will help you accomplish this task.

Triggering with a drum module

Certain drum modules, such as Ddrum or certain Alesis units, have the ability to trigger drum sounds based on an input signal that the module receives. This has been one of the earliest methods to accomplish sound triggering. Of course, there are issues of latency involved, but this can be corrected in the studio by playing back the track from the record head and delaying it so that it occurs early enough to compensate for the triggering time. With modern technology, triggering sounds this way is no longer an efficient method.

Non-real-time sound triggering

There are two methods of sound replacement for drums. The first is a non-real-time process, where you run a plug-in, such as Digidesign's SoundReplacer, which will go through and replace sounds by lining up the transients as performed by the drummer. You then have the option of creating a new audio track once all of the sounds have been replaced, or you can create an additional track to layer percussive sounds. This method has the advantage of not using up your computer's processing power. The disadvantage to non-real-time triggering is that you are not able to make changes to the triggered sound in real time. You will have to go back and retrigger the sounds again.

Real-time sound triggering

The other method of triggering drums in a recording is to do it via real time. The most popular application for accomplishing this in recent years has been through an application called Drumagog. This functions as a plug-in across the track that reads the transients in real time and triggers a selected sound. If your computer has the processing power to trigger sounds in real time, during a mix, then this would be the ideal method to use, as you have the ability to make adjustments to the sound, or change them completely without having to go through the triggering process again.

BUILDING A SOUND LIBRARY

Sampled drum libraries

There are numerous sample libraries available for use with drum triggering. Many drum triggering applications will come with a basic set of sounds for you to use right away. Otherwise, you can purchase many high-quality recorded drum sounds with many different variations. There are also many different drum libraries that can be found for free on the Internet from hobbyists. These can vary greatly in quality, but if you are persistent, you can find some good sounds to use.

Building your own sound library

Having your own library of sounds can give your studio and your projects some uniqueness that cannot be found elsewhere. You can take drum machines that you may have, and sample them to create a library of drum sounds. Additionally, you can record drums yourself to build a sound library.

When building your own sound library, it is best to keep track of the file format used by the drum triggering programs that you may be using. There can be many different file types used in digital audio workstations (DAWs), including WAV, AIFF, SDII, etc. You may need to do some file conversion in the future, but the most common format currently being utilized is the WAV file type.

Recording drums for triggering during a recording session

One of the easiest ways to build up your sound library is to record individual drum sounds when you are doing your basic tracking. If you are looking to create consistency to a drummer's sound via drum triggers, this will give you the ability to trigger the drummer's tracks, with its own drum sounds. This way you can keep the same sound, but have the advantage of a snare drum track without bleed from the other instruments.

Setting up your session for recording drum sounds

If you want to record the drummer's sounds so that you can trigger them at a later date, set up a session in which you are recording all of the drum tracks and nothing else. Even if you are looking to just have a kick and snare drum track to trigger, take the time to record every drum and cymbal several times.

Recording the drum sounds individually

While recording your individual drum sounds, make sure that you have the cleanest version possible of each of the drum hits. Get varying velocities of each drum sound. Drum triggering applications allow for triggering different sounds depending on the velocity received, so having a soft–snare hit trigger with a soft snare hit will put some of the feel back into triggered drum sounds.

If the drummer has different snare drums, record each of them so you can change up the sound in the future. Be sure to keep all of the microphones open while recording the drum sounds so that you can blend in some of the overhead or room microphone sounds into the drums if you choose.

Make sure that the sound has decayed completely after each hit. This is important to do with cymbals as well. Having the cymbals recorded is very handy, as there may be instances when the drummer may have had a great take, but missed a cymbal crash in a section where he or she wanted one. Simply pasting in one of these recorded cymbal hits will help salvage a great take that may have only one minor flaw. If there is bleed from the click track that was accidentally recorded on the last cymbal hit of the song, pasting in a previously recorded cymbal crash in place of the original version will help salvage the end of the song.

When recording the snare drum, get different versions of the drum with the snares turned on and the snares turned off. When you are recording anything else besides the snare drum, make sure that the snares are turned off, as they will rattle in resonance with all of the tom hits. This will prevent an unwanted rattle from being in your triggered drum sounds.

Capturing the snare bottom microphone

There may be times when you are working on a track that may not have a microphone on the bottom of the snare, but you may find that you would like to add one to the track through the use of triggering. You will find it advantageous to add different recordings of the snare bottom to your library for future usage.

Mixing the tracks for triggering

When it comes time to create these individual sounds for triggering, you may have more than one track on a single drum. Perhaps you have three different microphones on the kick drum that you will need to blend in to a single audio file to use for your trigger.

The first step is to edit the sound so that the triggered audio file starts with the drum sound without any silence in the beginning. Some drum triggering programs merely trigger an audio file, without regard to the content, so having the transient at the very beginning of the track to be triggered helps to create a more accurate drum-triggered sound.

Making sure that layered sounds are in phase with each other

If there is more than one microphone on a single drum, you need to be cautious to make sure that the sounds are in phase with each other. If one of the tracks is 180 degrees out of phase, then there will be frequency cancellation with the triggered drum sound. You can see this if the peak of one of the tracks is going down while the other tracks are going up.

In Figure 3.1, you will see that there are three microphones on the kick drum. The middle microphone is out of phase with the other two, and will therefore create phase cancellation. This is an easy fix, as there are other real-time plug-ins that will correct the phase. There may be an "invert" function, which will process the file and invert the phase.

Figure 3.2 has the same three tracks of audio, only the middle track has been processed with a phase-inversion function, which can be found in any DAW.

Adjusting the levels, processing, and fades for the triggered sound

In the same way that you mix all of the tracks for a song, you need to mix the multiple tracks for a single triggered sound. You want the final triggered sound to sound as close to the end result in the mix as possible. This includes equalizing the tracks, as well as finding the appropriate balance between each of them.

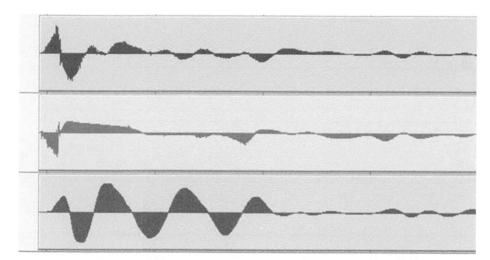

FIGURE 3.1
Three microphones
placed on the kick drum
track, with the second
track out of phase.

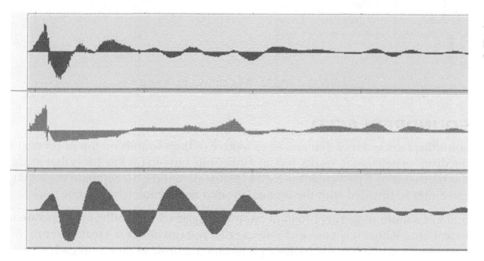

FIGURE 3.2
The second track has
had its phase corrected
with a plug-in.

You may find it necessary to apply a fade to the tracks individually before bouncing them down to create a single audio file. Figure 3.3 has a separate fade applied to each of the three microphones on the kick drum. This helps to tailor the decay of the sound, and you will be able to work with what you are looking for in the end result.

A note about using recorded triggered drum sounds for other projects

It is always wise to get the drummer's permission if you are planning on using his or her sounds on other people's work. It is a matter of common courtesy to ask them for permission so that you keep them satisfied with you in the future. You will find that a lot of drummers would be flattered if you wanted to use the sound of their drums for other projects.

FIGURE 3.3
The different tracks have been faded to tailor the sound for a triggering sample.

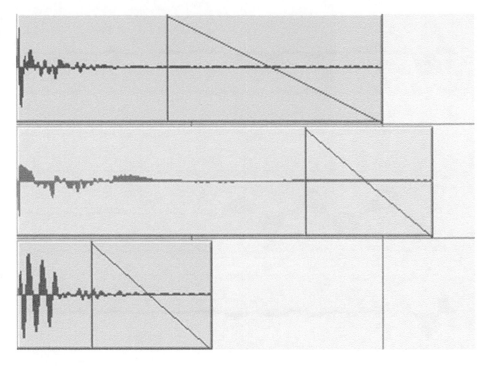

SOUNDREPLACER

SoundReplacer is a tool that is used to replace or layer sounds on top of preexisting drum or percussion tracks. It is an AudioSuite plug-in for Pro Tools that reads the transients of a particular track and creates an audio on a separate track that can either be blended with the original or used as a completely separate track.

It is not a drum triggering program, as drum triggers, historically, have operated in real time. Rather it is a piece of software that you can utilize to create an entirely new drum track. It literally pastes a prerecorded sample of audio on each of the transients.

The most common sounds that are used with SoundReplacer are the kick, snare, and toms. These sounds constitute the bulk of what we hear in a recorded drum track. They are the loudest sounds, so there will be fewer false transients from other instruments in the track that will create a false trigger.

Steps of SoundReplacer

The operation of SoundReplacer takes place in eight steps:

1. Editing the region you are going to use SoundReplacer across.
2. Selecting the destination track.
3. Loading one or more samples in SoundReplacer.
4. Adjusting the threshold of the sample(s).
5. Selecting the dynamics, peak align, and crossfade.

6. Processing the original track onto the new track.
7. Listening to make sure that everything triggered correctly.
8. Making editing adjustments to the newly triggered track.

Using SoundReplacer across a snare drum track

SoundReplacer, much like other triggering programs, will not work in every situation. This is particularly noticeable on a snare drum track that may contain ghost notes or flams, which may be difficult to trigger. That does not mean that SoundReplacer will not work on these tracks, but they may require specific editing so that the snare drum hits are triggered but the ghost notes remain in the original track.

Preparing edits on a track before using SoundReplacer

SoundReplacer is something best added to the session after all drum timing edits and corrections have been made. If you are looking to correct the timing of the drums manually or through the use of Beat Detective, it is advised to complete these edits cleanly before utilizing SoundReplacer.

CLEANING UP A TRACK BEFORE APPLYING SOUNDREPLACER

Knowing that SoundReplacer specifically looks for transients, you can make the process more efficient by eliminating the background noise in the track, so there will be fewer false transients that can trigger false hits in the new drum track that you are creating.

USING STRIP SILENCE TO CLEAN UP BACKGROUND NOISE

You can eliminate much of the background noise by applying either a gate, and writing it to the track, or by utilizing the Strip Silence feature of Pro Tools, which functions similarly to a gate. The main difference is that there is no attack or decay functionality; it merely eliminates regions between transients. This works well for preparing a track for SoundReplacer as you also have the ability of padding the start and end of the transient by a specified number of milliseconds.

Before you utilize Strip Silence, duplicate the playlist of the track that you are going to be editing. This again gives the ability to go back to a previous version with ease in case you change your mind about using SoundReplacer in the future.

For this example, we are going to be using SoundReplacer across a snare drum track. As you can see in Figure 3.4, there are some other drum hits in the background of the snare drum track. The high peaks are the transients from the snare drum hit, but the information in between is bled from the other microphones, particularly from the kick drum.

If you add compression to the snare drum when you are mixing, you will find that you are increasing the background noise. Some of this noise can be eliminated by placing a gate across the track. However, having a cleanly triggered track will have no background bleed and no additional background noise, no matter how much compression you apply.

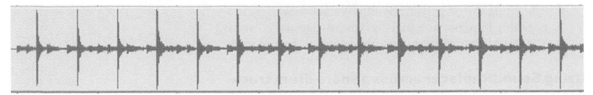

FIGURE 3.4
The snare drum track with visible background bleed.

Double-click into the region that you are looking to clean up. After you open up Strip Silence from the Edit menu, you will see that the first option is Strip Threshold. Before adjusting this, reduce the Min Strip Duration to 0-ms. This will prevent Strip Silence from missing any of the gaps between transients due to their duration being too close together. Next, go to the Strip Threshold slider. Start with this fader all the way up, and lower it until all of the transients are indicated with a white line in the region, but no further, as you risk Strip Silence including unwanted transients into the edits.

Adjust the Region Start Pad to 1 or 2 ms, as this will allow any sound from the snare drum to remain as it builds up to the transient, without stripping away any of the initial attack. Adjust the Region End Pad so that it is long enough to contain most of the decay of the snare drum sound, but stops short of any other transients. When you adjust these settings, you will see that what originally looked like a white line was a small rectangle indicating the areas in the region that will not be stripped away.

As you can see in Figure 3.5, the Region End Pad of 215-ms has kept most of the decay, but there are no false transients for SoundReplacer to create a false trigger.

After you have dialed in the appropriate settings, click "Strip," and you will see everything that is not enclosed by the white rectangles stripped away. After this has been completed across the track (Figure 3.6), you are ready to start using SoundReplacer.

FIGURE 3.5
The Strip Silence feature, indicating the regions it will retain with white boxes.

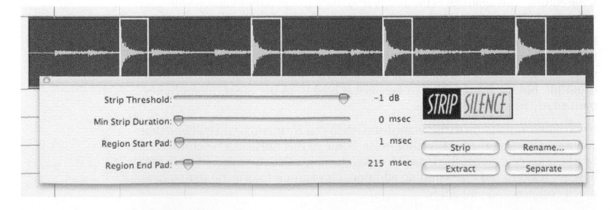

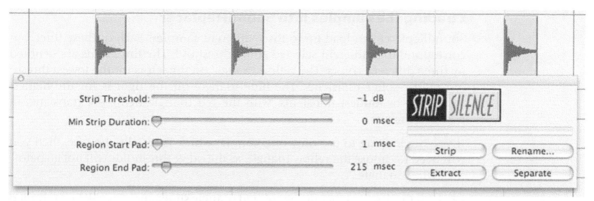

Creating a track for SoundReplacer

With SoundReplacer, you have the ability to either completely replace the sound or layer the recorded sound with a triggered sound. The best method to accomplish either of these is to place all of the triggered sounds onto a new track. You may find that you need to work with SoundReplacer's accuracy, either manually or through the use of Beat Detective, and so having both the original track and SoundReplacer's track next to each other will come in handy.

Begin by creating a new track and labeling it something like "Snare Trigger," so as to make it stand apart from the other tracks.

SELECTING THE DESTINATION OUTPUT FOR SOUNDREPLACER

With SoundReplacer open, you have the ability to select the destination track for the newly triggered sounds. If there is not a destination track selected, then SoundReplacer will function as a normal AudioSuite plug-in. There is no disadvantage to having the triggered sound go to a different track, with the exception of adding another voice to the track count in your digital audio workstation. Going to a different track will allow you to blend the two as well as compare the timing. You can select a destination track at any point in time before you process the audio, but it is best to get into the habit of doing this first so you do not forget later.

Sample rate and SoundReplacer

SoundReplacer will work with any session sample rate up to 192 kHz. However, you will need to make sure that the sounds you are loading into SoundReplacer are at the same sample rate as your current session. If you are loading samples that are at a higher sample rate than your current session, SoundReplacer will not do any conversion, and the samples that you load into SoundReplacer will be pitched down. The sounds will be pitched up if they are at a lower sample rate.

If you are keeping a library of samples to use with SoundReplacer, it may be best to keep different versions of each sample rate using all of the different sample rates possible.

FIGURE 3.6
The final stripped version, clean of most of the background noise.

Loading the samples into SoundReplacer

SoundReplacer can load up to three different samples. Each of these three can correspond to a different selected audio threshold. The thresholds are denoted by three different colors. The yellow triangle on the left is for the lowest threshold or the softer transients. The blue triangle on the right is for the highest threshold, or loudest transients, with the red triangle being the transients in between the two.

If you are looking to use only one sample, which is usually the case, then you will need to choose the yellow triangle, as the other thresholds will not go below the yellow triangle.

Click on the disk icon below the colored triangle slider and you will be prompted to load a sample from a hard drive. Once the sample is loaded you can begin to lower the threshold.

Adjusting the threshold for the triggered sounds

Before you begin to lower the threshold, you need to see the sounds in SoundReplacer's audio window. You can accomplish this by clicking "update." There is an "auto update" button that will update the window anytime you select a different region; however, if you click out of the region you will get an error message saying "No audio was selected," in which you will have to make another selection and press "update." This can be tedious and annoying, so it may work better for you to leave "auto update" unselected.

There is a circular zoom button in the middle of SoundReplacer, which will allow you to increase the visual amplitude of the audio in the window by clicking on the up arrow. The left and right arrows will allow for horizontal zooming of the audio in the window. You can zoom all the way out so you can see all of the transients; however, if it gets too crowded, you can just scroll through to make sure that all of the audio is being selected.

As you lower the threshold of the yellow triangle slider, you will see SoundReplacer draw yellow lines to correspond to the velocity triggering. This indicates where SoundReplacer is going to paste the samples in the selected destination track.

In Figure 3.7, you can see that there should not be any false triggers due to the editing of the region prior to using SoundReplacer. With a dramatic lowering of the threshold, very low-level transients in SoundReplacer can be detected as seen in Figure 3.8.

Crossfade

There is a button in SoundReplacer that you can select if you decide to place a crossfade at the end of the sample, right where the subsequent triggered sample begins. This will help keep the decay of the drum from being cut off right at the beginning of the next attack. With crossfade enabled, there is the potential for phase cancellation if the triggers are too close to each other.

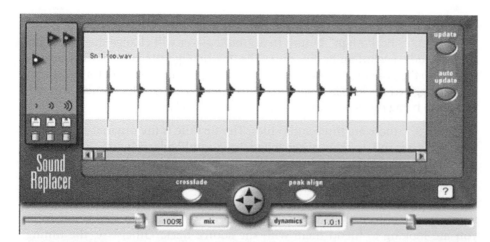

FIGURE 3.7
The detected transients being indicated in SoundReplacer.

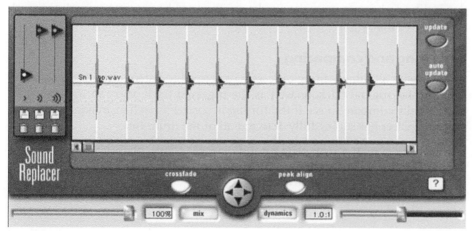

FIGURE 3.8
With the threshold lowered, SoundReplacer has detected a false transient.

Dynamics

SoundReplacer will match the velocity of the incoming material. If you are looking to add more consistency to the triggered sounds, or increase their variation, then the dynamics adjustment will create your desired effect.

There is the option of controlling the dynamic range of SoundReplacer with the dynamics option. This function is similar to a compressor or expander. If you think of these ratios in terms of compression and expansion, the ratios used by SoundReplacer are the opposite of what you might think they should be. Compressor ratios are 4:1, 2:1, etc., whereas the same dynamic compression ratios in SoundReplacer are listed as 1:4, 1:2, etc. The same holds true for the expansion in SoundReplacer.

Contemporary pop records have a more consistent sound to the drums, so you may find yourself only looking at the compression side of the dynamic range adjustment. If you are looking for no dynamics changes, you can press the dynamics button, and it will go from green to gray. This will then paste the exact same hit at the same velocity for every trigger.

"Peak Align"

Enabling the "Peak Align" button in SoundReplacer will change the way that it triggers audio according to the source file. It will align the peak of the triggered sample to the peak of the source audio. This may or may not work for the timing of the triggered track. It is best to try both versions to see which one ultimately winds up being the closest in time to the source track.

In Figure 3.9, there are three audio files. The first file is the original snare drum track. The second track was generated by SoundReplacer with "Peak Align" turned off, and the third track with "Peak Align" turned on.

You can see with the track that did not have "Peak Align" enabled (Figure 3.9b) that it was much closer in time to the original source track using SoundReplacer. Sometimes tracks generated by SoundReplacer will require a slight nudging if you are looking to blend them with the source track. Otherwise, the timing, depending on how accurate the results are, may not be accurate enough to stand alone in the mix.

Listening and comparing

Just like with all other editing processes, you need to listen to the end result. By keeping the original track, as well as the triggered track, you can then audibly and visually compare to see if the timing is correct as well as making sure that SoundReplacer has successfully detected all of the drum hits.

SoundReplacer may add an extra hit if it detects a transient, or it may miss a few depending on the consistency of the drummer, as well as the amount of bleed of the other drums into the microphone.

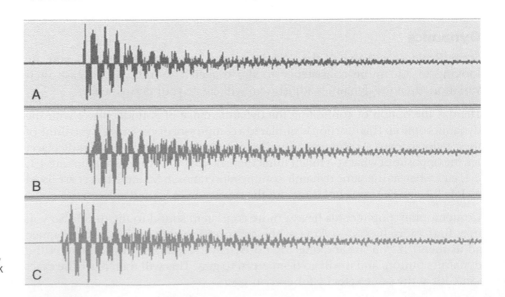

FIGURE 3.9
(a) The original snare track, (b) track with "Peak Align" turned off, and (c) track with "Peak Align" turned on.

Using SoundReplacer on tracks that have ghost notes

If you wish to use SoundReplacer across tracks that have ghost notes, careful editing must be done in order to separate out the tracks that you wish to trigger the snare drum. You can try to use SoundReplacer across those tracks and hope that the threshold triggers correctly, but an inconsistent drummer may lead you to spend more time retriggering false transients than if you manually edited the snare drum track before using SoundReplacer.

In Figure 3.10, you will see a snare drum track that has ghost notes played on the snare in between strong snare hits. The ghost notes are impossible to trigger, and they would not sound right, even if they were easy to trigger, as the variety of the hits are what make them sound like ghost notes.

The first step in making edits to tracks that contain ghost notes is to create a new track with which to put the edited ghost notes on. This will separate out the strong transients that you want to trigger from the ghost notes that you do not.

Placing edits in the source region, in between all of the hits of the snare and the ghost notes, will make it easy to place the ghost note regions into the lower track. You can easily do this by zooming and placing the selector cursor right where you are looking to edit. Then just press "B" with single-key shortcuts enabled, and Pro Tools will put an edit in the region where the cursor is located.

In Figure 3.11, you will see the edits in the region in between all of the ghost notes and the main drum hits. From here, just use the Grabber tool and drag the edited regions down to the newly created track below the current snare drum region. Now you will have two new tracks, as seen in Figure 3.12.

From here, the main snare track is edited enough so that you do not need to use Strip Silence for any further editing. You can proceed to use SoundReplacer to trigger the main snare hits on a new triggered track, as mentioned previously.

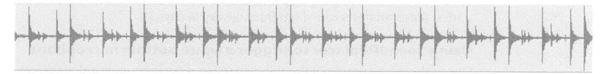

FIGURE 3.10
A snare drum track with ghost notes that are usually impossible to trigger.

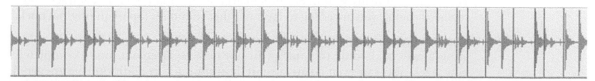

FIGURE 3.11
The track with region breaks placed between the snare hits and the ghost notes.

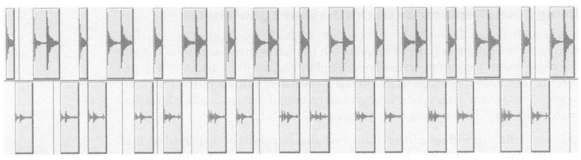

FIGURE 3.12
The separated ghost notes have been dragged into their own separate track.

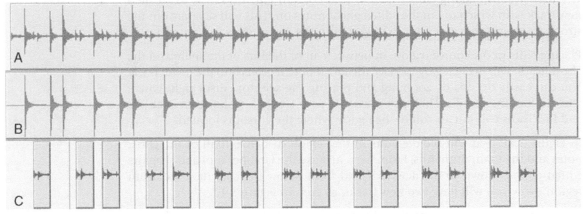

FIGURE 3.13
(a) The original snare track, (b) the triggered snare hits, and (c) the ghost notes.

In Figure 3.13, you will see three tracks. The first is the original unedited snare drum track, the second track is the triggered main hits of the snare, and the third track is the ghost notes edited out from the original track.

Using SoundReplacer to trigger a snare bottom microphone

One of the nice things about the ability to trigger different sounds is that you can add sounds that may not be in your original session. If you are mixing a track that you have gotten from a client, and during the mix you wish there were a microphone on the bottom of the snare drum, you now have the ability to trigger one. This works out really well in SoundReplacer. Rather than have a new track and a replaced sound, you are going to layer another sound on top of an original track.

Run through the steps previously described in this section, as if you were creating a new track, only this track will not be a replacement, but a completely new track. You can even create a snare bottom track in addition to a triggered snare top track. For the sake of having a more accurate time alignment, it is recommended that you use the original track when creating both new tracks with

SoundReplacer, rather than trigger the bottom snare microphone with the triggered snare top microphone.

A trick for creating phase-coherent snare top and bottom tracks

Since there may be slight timing differences when using SoundReplacer across different tracks, creating multiple triggered sounds from the same source can be tricky. A simple remedy for this is to create a stereo drum sample, with the snare top on the left and the snare bottom on the right. You can then trigger this stereo sample with the original track and create two new phase-coherent snare top and bottom tracks at once.

Since SoundReplacer will not trigger a stereo sample from a mono source, you will need to fool it into thinking that you are working with a stereo track. To do this, you can duplicate the original track from the Track menu (Figure 3.14).

From this point you need to create a stereo track and drag both regions into it. This will create a stereo track with identical audio regions in both the left and right sides, as seen in Figure 3.15.

Create a stereo destination track from the stereo source, and now SoundReplacer will function as before, with the newly triggered snare samples being the left and right sides of the new track.

You then need to make this stereo track into a dual-mono track, which is easy to do in Pro Tools. Just highlight the stereo track and go to the Track menu in

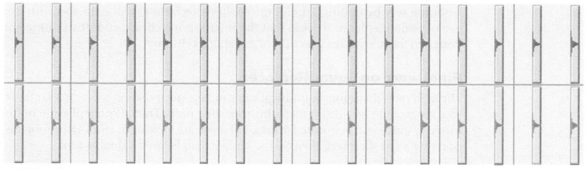

FIGURE 3.14
The edited single snare track has been duplicated.

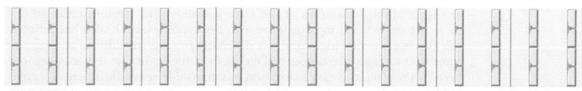

FIGURE 3.15
The mono, duplicated snare tracks have been dragged into a single stereo track.

FIGURE 3.16
A triggered snare top and bottom track created from a single snare track.

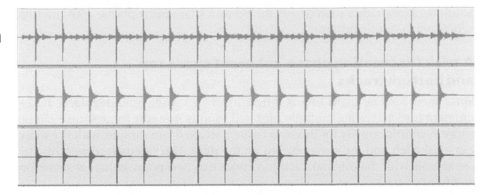

Pro Tools. From there, select "Split Into Mono," and Pro Tools will create two new mono tracks from the original stereo track. You can then delete the stereo-triggered track and work with the two mono tracks after you have panned them both back into the center. In Figure 3.16, you can see the two newly triggered snare top and bottom tracks created by running SoundReplacer only once. These are shown beneath the original snare drum track.

Other drums triggered with SoundReplacer

SoundReplacer works equally as well with toms and kick drums as it does with snare drums. Just follow the same procedures mentioned above for each of these different instruments, and you can replace any struck drum, as long as the transients of that drum are louder than the transients from the other elements that may bleed into that track. Even if you plan on using SoundReplacer from the very beginning, careful microphone technique will help you with the use of SoundReplacer. Be sure that the microphone choice and positioning are designed for maximum rejection of extraneous drums.

Final word on SoundReplacer

Through careful editing, SoundReplacer can become a session saver or a fantastic tool for creating new drum tracks from existing parts. There is no replacement for carefully tuned and performed drums, but if you are looking to eliminate problems or create a specific effect, then SoundReplacer can be an invaluable tool.

DRUMAGOG

Drumagog is a plug-in that will replace sounds in real time. It has the ability to trigger multiple samples of the same instrument in random order so that there is some natural inconsistency with the drum sound. It also has different velocity layers with which to trigger different samples. In addition to triggering sounds, Drumagog can output MIDI data from the input that it receives as well. Because the plug-in is easy to use, and has many different adjustments for creating triggered sounds, Drumagog has become one of the most-often-used plug-ins for drum triggering. It is available on both Mac and PC platforms, and it is supported by a variety of host DAWs. As with many complex plug-ins, there are a variety of versions to suit the needs and budget of the engineer, as well as an upgrade path for the different versions.

Applying Drumagog across a track

Once installed, Drumagog will have different versions of the plug-in. Depending on your DAW, and whether or not it supports automatic delay compensation, the plug-in instance should be based on what is recommended by the manual. There is a version that will have a stereo output from a mono source or a mono output from a mono source. For standard drum replacement, such as kick, snare, or toms, the mono output should be used depending on the samples you are wishing to trigger.

The Drumagog plug-in should be placed as the first plug-in of the track, unless you are using gating or equalization plug-ins to clean up the sound prior to the Drumagog plug-in.

Hierarchy of Drumagog files

The main file format used by Drumagog is the GOG format. This format contains all the samples, which includes the dynamic groups and their thresholds, as well as which samples are assigned to different groups. GOG files do not contain any of the settings from the Main or Advance tabs of the Drumagog plug-in.

When opening up Drumagog you will see categories for each of these GOG files. Each category may have any number of GOG files. Inside of the categories are each of the GOG files that, when selected, will load the appropriate samples and their settings. These categories and files correspond to the Drumagog location on your hard drive. The categories are folders inside of the GOG files folder, with the GOG files placed inside the corresponding category folder. See Figures 3.17 through 3.19.

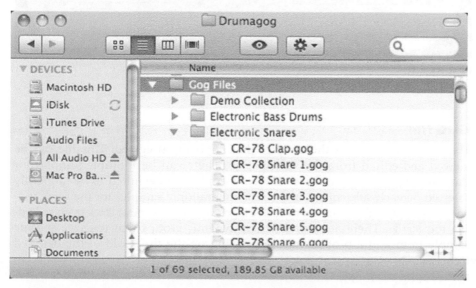

FIGURE 3.17
The GOG files folder in the Drumagog application folder where all the GOG files are stored.

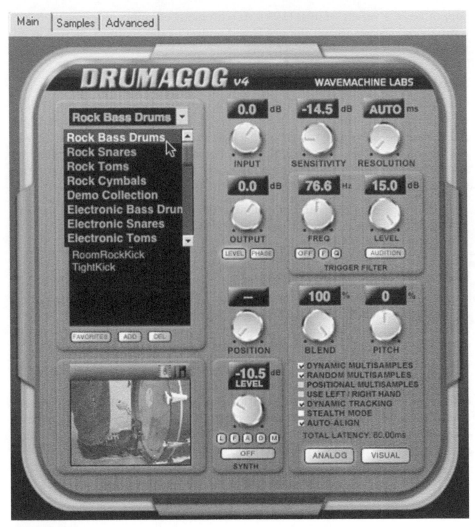

FIGURE 3.18
The different categories of drum sounds, derived from the GOG files folder.

GOG file

The GOG file contains the samples that are being triggered. This file can be viewed and edited from the Samples tab. There can be multiple samples for the same triggered track; these different samples are defaulted to be randomly selected on each subsequent trigger. Having multiple samples for the same trigger can have a slight variation to each trigger, which prevents a track from sounding too sterile. There are also different dynamic groups that trigger a specific sample from a group of samples depending on the threshold of the detected transient. These groups are color coded so that they can be visually identified in the Samples tab (Figure 3.20).

Editing tracks

As with any drum triggering application, the best results will be achieved by editing the tracks to remove as much unwanted background noise as possible. With a consistent drummer, just utilizing the sensitivity adjustment in the Drumagog plug-in can be sufficient for most tracks. Certain tracks will have more background bleed than the softest transient, which will require individual editing across that track (Figure 3.21). This is common in instances when there are dynamic builds in the drum part or poor microphone technique.

Since Drumagog functions in real time, edits to the track can be made as you are working, whenever you hear a missed or false trigger.

Adjusting the sensitivity

After selecting the appropriate GOG file for the track that you are working on, just play the track and listen to the accuracy of Drumagog. The sensitivity, which can be thought of as the transient threshold, can be adjusted from the Main tab (Figure 3.22). To more accurately adjust the sensitivity, click on the "Visual" button

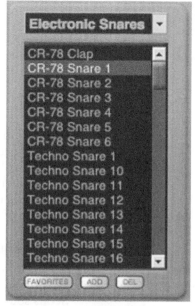

FIGURE 3.19
The individual GOG files, located inside of the folder categories.

FIGURE 3.20
The individual samples in the GOG file, with the velocity layers indicated by color.

FIGURE 3.21
A tom track, with soft hits, where the surround tracks are muted to eliminate bleed.

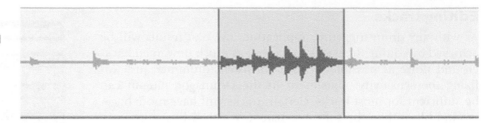

at the bottom of the Main tab. This visual window will show the triggering done by Drumagog and place white circles where it is triggering the sounds. The sensitivity can be adjusted from the visual window by manually raising or lowering the sensitivity triangle with the mouse (Figure 3.23). This way you can see where Drumagog may be missing some of the transients because the sensitivity is set too high. If there are false transients, you can raise the sensitivity.

Using equalization and filters to enhance accuracy

Because Drumagog is looking for a specific transient for the sounds that it triggers, filters and equalizers can be applied to help eliminate some of the extraneous sounds, as well as apply a boost to specific harmonic frequencies of the instrument to aid in triggering. Drumagog has filters built in, which will allow you to achieve more accurate triggering, especially in cases where there is a broad dynamic range.

For a snare track, there may be bleed picked up from the kick drum. A high-pass filter can be engaged to eliminate most of the frequencies picked up from the kick drum (Figure 3.24). You have the ability to adjust the cutoff frequency as well as the amount of attenuation. Clicking on the "Audition" button will bypass the triggering of Drumagog and allow you to hear how the filtering is affecting the source track. This helps to dial in the filtering to allow the maximum amount of transient with a minimum amount of bleed.

When applying a filter to a tom track, you can use the boosting equalizer of the filter to accentuate the attack of the tom. This will help increase the accuracy during the dynamic sections of the song.

FIGURE 3.22
The analog sensitivity knob to adjust the triggering threshold.

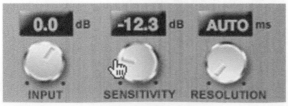

Any filtering done to the source track, whether it is a high-pass filter or the boosting of a particular frequency, will affect the overall input level from the source track. Any high-pass or low-pass filtering done will result in a softer input signal. If a frequency is boosted with the parametric equalizer, then the overall input signal will be amplified. These filterings can affect the way that the sound is triggered in Drumagog. The difference in gain can be compensated by adjusting the input level of the source track up or down depending on the filtering done. After any compensation has been added to the input signal, the sensitivity may still need to be adjusted, depending on the triggering of the track.

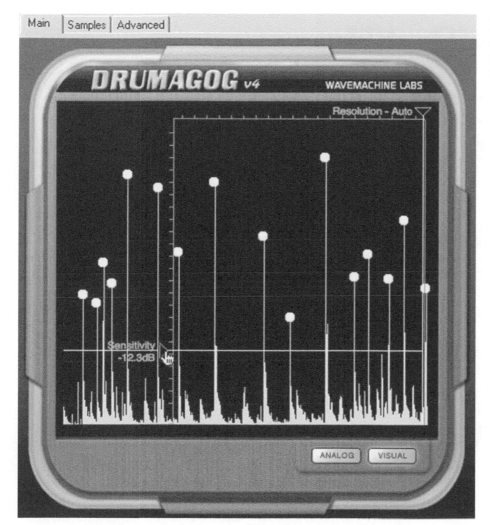

FIGURE 3.23
The visual screen,
where the sensitivity can
be manually adjusted
according to the input.

Fine-tuning the alignment of the triggered sound

Clicking on the Advanced tab in Drumagog will allow you to fine-tune the triggering of samples. The Auto-align section under the Advanced tab will allow you to select the type of triggering to be done. There are two types of styles for the triggering. The first is the psychoacoustical triggering, which is the default (Figure 3.25). This mode works best in most instances. The actual peak setting will align the peak of the incoming source file with the peak of the triggered sample. This mode will work best if there are any phasing issues with the triggered sound, or if you are using samples that you have created yourself. This will compensate for any inaccuracies in the initial attack of your own sample.

FIGURE 3.24
Selecting the high-pass filter for the source input of the trigger.

FIGURE 3.25
Selecting the method of alignment for Drumagog.

Controlling the dynamic of the triggered sound

As with other triggering programs, you have the ability to control the dynamics of the performance. Drumagog defaults to maintaining 100 percent of the dynamic range. This will keep the soft parts soft and the loud parts loud. If you are wanting to have more consistency to the sound, reduce the dynamic tracking, which will make each hit more consistent with the lower setting (Figure 3.26). A setting of zero will have the same dynamic output for every hit of that drum track.

Dynamics can also be adjusted in the GOG file. There are different layers of dynamics within the GOG file. The GOG file can have different groups of samples, each set for a specific dynamic threshold. Clicking on the "Samples" tab will allow you to edit the dynamics of the different groups. Selecting the "Groups" tab within the Samples window will show you the different layers of dynamics for the different samples.

FIGURE 3.26
The dynamic tracking adjusts the consistency of the triggered sounds.

The level of the transient, inside the Groups tab, will determine which group of samples is triggered, according to the threshold set. If the drummer is inconsistent, his or her performance may be triggering samples that are designated for softer hits. The threshold of the different dynamic groups can be manually edited by lowering them with the mouse. This will make sure that the intended samples are triggered at the appropriate thresholds. If you would rather have only one dynamic group that can be triggered, the additional layers can be deleted inside the Groups window by selecting that group and clicking "Delete" at the bottom of the window. It should be noted that any editing within this Samples window also makes changes to the GOG file that has been loaded. These settings can then be saved into a different GOG file. Click into the "Groups" tab inside the Samples window and enter in a different name at the top of the window, and click the "Save GOG" button. See Figures 3.27 and 3.28.

FIGURE 3.27
Saving the GOG file
by double-clicking the
name and pressing
"Save GOG."

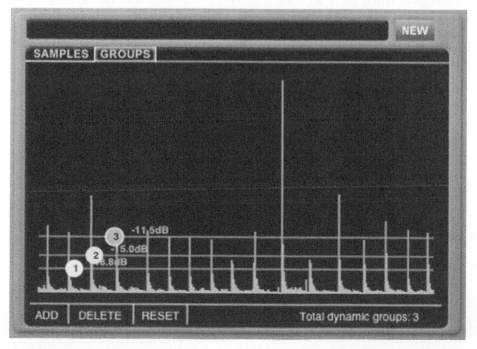

FIGURE 3.28
Three different color-
coded dynamic groups,
which can be used to
trigger different samples.

Triggering the synthesizer to create deeper sounds

One of the added features to the platinum and professional versions of Drumagog is the synthesizer. This can be used to trigger an additional synthesized sound on top of the sample. This is similar to the old analog console trick of gating a sine wave oscillator with the input from the kick drum. When set to a low frequency this adds an extra low harmonic to the kick drum sound.

The synthesizer is accessed from the Main tab. It should be set for a sine wave, as any other tone will be noticeable as a separate, pitched sound. You have the ability to select the frequency, level, attack and decay times, and the mix of the synthesizer with the samples. The frequency should be set low enough so that the pitch does not sound like a separate instrument. A sine wave between 20 Hz and 40 Hz will create a tone low enough to add a subfrequency to the kick drum. The level of the sine wave can then be increased to the point where it contributes a noticeable low-frequency addition to the kick drum. In addition to adding it to the kick drum, it can also be added to the floor tom for a similar effect.

Improving transient detection

Even with optimal settings, transient detection can only work as well as the source material provides. There are the instances when the bleed from the other drums is louder than the actual drum being triggered. In other instances, there can be dynamic builds with very soft transients that require some extra work so that the transients are accurately detected.

In instances when there are false triggers, the solution is to merely delete the source of the false trigger from the track. This will prevent you from having to adjust the sensitivity settings in Drumagog, if everything else is being triggered appropriately.

For sections when there are dynamic builds in the drum tracks, it may be best to create a whole separate instance of Drumagog for these sections (Figure 3.30). This will allow for separate settings in Drumagog while retaining the triggering of the rest of the track.

FIGURE 3.29
Adjusting the frequency of the sine wave synthesizer to add an extra low harmonic to the kick drum.

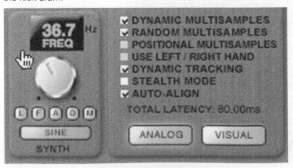

A more drastic equalizer can also be applied prior to the Drumagog plug-in, in order to more accurately fine-tune the tone and the transients of the source track. Since Drumagog only allows for a single equalizer to be placed across the source track to increase the accuracy of the triggering, any parametric equalizer can allow for multiple bands of equalization prior to triggering. Focusing in on the attack and main fundamental of the source track while reducing any unnecessary sound of the drum will help (Figure 3.31). This equalizer does not need to sound or look good; it is merely there to increase the frequencies of the transients.

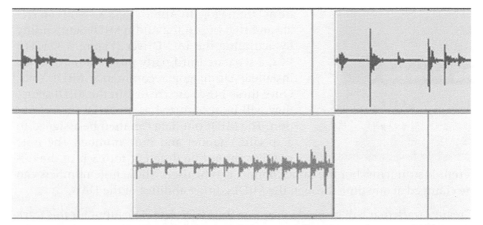

FIGURE 3.30
A dynamic build, separated onto a second track for triggering with different settings.

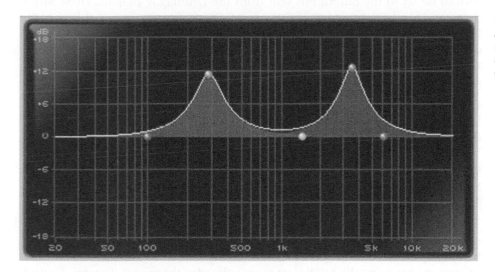

FIGURE 3.31
A dramatic equalizer plug-in to accentuate the attack and fundamental of a tom track.

Exporting MIDI data from Drumagog

Drumagog has the ability to output the transients that it detects as MIDI data. This can be used to trigger additional percussion sounds from separate MIDI modules or software instruments. Due to the nature of MIDI, the data are not as accurate as the actual triggering, but they still can be used with some careful editing and timing correction.

On the Advanced tab in Drumagog, there is the ability to both send and receive MIDI data. Selecting "MIDI Out Enable" will give you the option of selecting which MIDI port to send the triggered data. This MIDI data can then be recorded on a separate MIDI track. You also have the ability to select the MIDI channel and note number of the data sent from Drumagog.

Drumagog is not recognized as an MIDI output device since it is an instrument. You will need to create a separate MIDI bus to capture the MIDI data

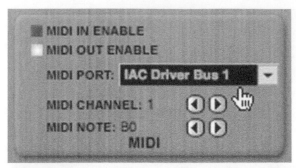

FIGURE 3.32
Selecting the MIDI destination for the output of the detected triggers.

from Drumagog. In Apple's "OS X," you can create an MIDI bus in the Audio MIDI Setup utility by activating the IAC Driver (Figure 3.32). For PCs, a separate third-party program needs to be installed. Drumagog recommends MIDI Yoke. Once these buses are created in the MIDI setup, they will be recognized as an MIDI in and out port. The MIDI out data can then be assigned to a specific channel and note number. The note number should be different for each of the different instruments, but the same channel can be used. These note numbers can be changed at any time through the MIDI editing abilities of the DAW.

Create a track that will be used to capture the MIDI data. The input for this track should be set to the output bus sent from the Drumagog plug-in. These MIDI data can then be recorded as if they were coming from an external keyboard. If there are multiple instances of Drumagog across different tracks, these MIDI data can be captured on a separate track for each instance. They can be either recorded on a single track for drums or recorded on individual tracks. It will be easier to adjust the MIDI data if they are recorded on separate tracks.

Drumagog will output the MIDI data with velocity information based on the level of the transient. This varying velocity does not change, regardless of whether or not the Dynamic Tracking is set to 0 percent for a consistent output. If you are looking to have a consistent MIDI trigger, then you will need to adjust the MIDI data in your DAW so that the velocity is consistent. With the MIDI data captured, you can then use a whole host of MIDI editing features that are inside your DAW software (Figure 3.33).

There can be many reasons why the MIDI data are not 100 percent accurate with the performance of the drums. MIDI is, in itself, a sloppy format. There can also be latency involved in the system, when it may take a few milliseconds for the instrument receiving the MIDI data to actually play the sound. To work around this timing issue, the MIDI data can be shifted earlier in time to accommodate the latency inherent in an MIDI system. Many DAWs allow you to set an MIDI offset without having to nudge the region earlier.

Using your own samples for triggering

Drumagog allows you to use your own samples as well as any downloaded or purchased GOG files. It gives you the tools for creating your own GOG files from samples, by either importing the sounds, manually, or recording them directly from a track.

In order to keep your GOG files organized, it is best to begin creating your own category of sounds. To create a category that will be recognized by Drumagog, go to the location on your hard drive where the GOG files are stored and create a new folder among the other categories (Figure 3.34). This folder will then

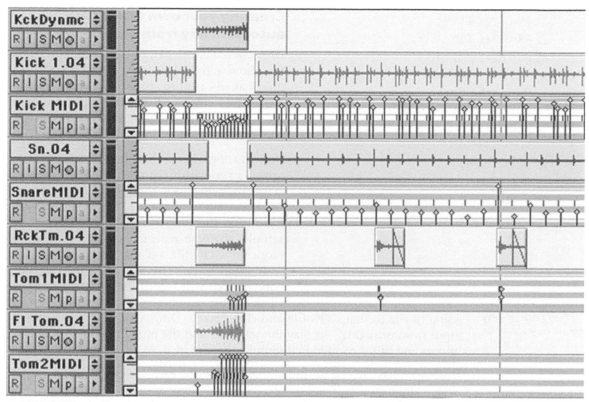

FIGURE 3.33
MIDI tracks recorded from Drumagog placed beneath the source audio tracks.

show up when opening up the Drumagog plug-in. If the Drumagog plug-in is already open, you will need to reopen that instance in order for it to recognize the new folder.

To begin loading in your own samples, click on the "Samples" tab and then click "New." Double-click the name, which will save the new GOG file in the categories folder. You can then select "Add From File," which will then bring up your operating system's finder. Even if your sample is at a different sample rate than your session, Drumagog will convert the sample inside the plug-in by selecting that option under the Advanced tab.

If you are only using one sample, under the Groups window, right-click at the bottom of the screen and select "one dynamic level." If adding multiple layers of samples, you can select "multiple dynamic levels." See Figure 3.35.

After loading in all samples, you can select which dynamic group to put the samples in. Simply click on the sample at the bottom of the plug-in window. You have the choice as to which dynamic group to place the sample (Figure 3.36). The dynamic groups are color coded in the window, which matches the threshold when switching to the Groups tab.

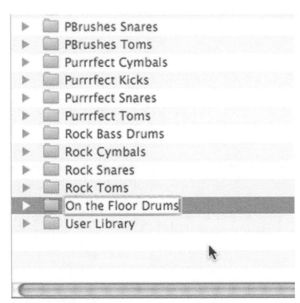

▶ 📁 PBrushes Snares
▶ 📁 PBrushes Toms
▶ 📁 Purrrfect Cymbals
▶ 📁 Purrrfect Kicks
▶ 📁 Purrrfect Snares
▶ 📁 Purrrfect Toms
▶ 📁 Rock Bass Drums
▶ 📁 Rock Cymbals
▶ 📁 Rock Snares
▶ 📁 Rock Toms
▶ 📁 On the Floor Drums
▶ 📁 User Library

FIGURE 3.34
Creating a new category that will be visible inside the Drumagog plug-in.

Creating your own GOG files automatically from a track

Drumagog makes it easy to create your own GOG files from a recorded drum track. It has a feature that will read in consecutive samples from a track and create individual samples inside of the GOG file. These samples do not need to be edited, as the plug-in does all the work for you.

Begin by opening up a session that has the drum samples you wish to capture. If the samples are across only one track, then open up an instance of Drumagog on that track. Create a new GOG file inside the plug-in. Highlight the number samples that you wish to capture on the track. From the GOG file, on the Samples tab, select "Add From Track." This will bring up a separate window that will record the different samples. Click on the number of samples that you are going to add to that GOG file and click "Start." Drumagog will be waiting for input from your DAW, so play the selection and the plug-in will automatically import and edit the individual samples. See Figure 3.37.

Drumagog has created an individual sample for each hit of the drum (Figure 3.38). Drumagog will automatically assign these samples a dynamic group based on their input level. These dynamic groups can be changed manually.

If there are multiple tracks for a single instrument, such as a kick or snare drum, then you will need to bus those instruments into a single output (Figure 3.39). Place the Drumagog plug-in across the bus. Adjust the balance between the

FIGURE 3.35
Adjusting the number of dynamic levels for the GOG file.

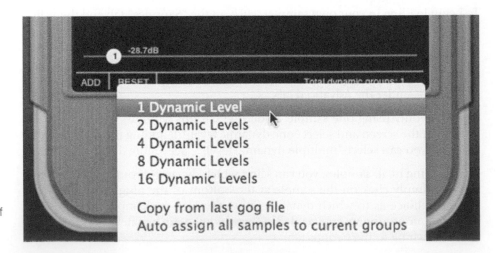

-28.7dB

ADD   RESET   Total dynamic groups: 1

1 Dynamic Level
2 Dynamic Levels
4 Dynamic Levels
8 Dynamic Levels
16 Dynamic Levels

Copy from last gog file
Auto assign all samples to current groups

FIGURE 3.36
Adjusting the dynamic group of the imported samples.

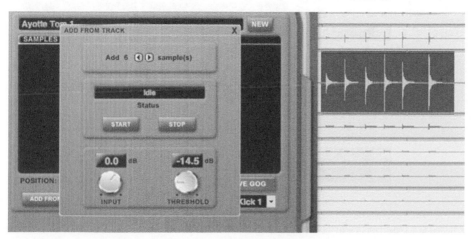

FIGURE 3.37
Selecting Drumagog to record six consecutive drum hits to create individual samples.

FIGURE 3.38
The six different drum samples that have automatically been recorded and edited.

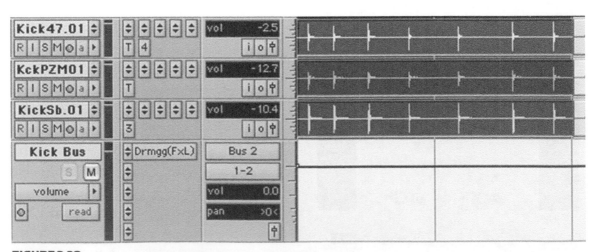

FIGURE 3.39
Three tracks from a kick drum, bused to a single output with the Drumagog plug-in for recording.

different tracks, as well as any desired equalization, and then record the output of the bus as you would a mono track in Drumagog.

A FINAL WORD ABOUT REPLACING OR TRIGGERING SOUNDS

Once any sound triggering or replacing has been done, make sure that the tracks are improving toward the ultimate goal. If you are looking to create a very consistent sound, then the sounds should not have very many dynamic layers. If you are trying to retain a natural feel to the drums, make sure that they do not have the "machine gun" effect that can come from triggered sounds.

CHAPTER 4
Pitch Correction

PITCH CORRECTION OVERVIEW

Being able to correct a pitch in the studio has been a recent technological advancement. Pitch correction entails being able to identify the pitch, determine the correct pitch, and process the track with the pitch correction so that it does not sound artificial.

Since the beginning of digital audio workstations (DAWs), there have been techniques to adjust the pitch of tracks. This pitch adjustment would adjust the pitch of the entire audio region. This was not the best algorithm for adjusting the pitch of a lead vocal, as the result would have an artificial sound.

With the advance of processing and technology, pitch correction evolved into algorithms that were able to identify and correct the pitch with minimal artifacts. This pitch correction is able to conform identifiable pitches to a predetermined scale. This was the beginning of the modern era of perfectly intonated vocals.

In the past few years, having a vocal track on a pop record with slight fluctuations in pitch has become rare. Since correcting pitch in the studio has become easy to do, the standard has become to apply it almost all the time. This has led to a sterility in the sound of vocals on a pop recording. Much of the emotion carried by the vocalist is through the subtle variations in pitch.

For years, pitch correction has only worked across a monophonic track. This is because it is easier to identify the pitch from a monophonic sound source, as it has a periodic waveform. This periodic waveform is easier to recreate into a different pitch. More recent advancements have allowed for pitch adjustment beyond that of the monophonic instrument. This involves a much more complex process of identifying the fundamentals and harmonics of individual notes on a track.

There is constant debate over whether or not pitch correction has helped or hurt the quality of recordings. On one hand, some of the artistry and work that vocalists go through to hone their craft has been eliminated. They no longer need to record every single line as accurately as possible. However, more emphasis has been placed on getting the best feeling and inflection in the voice, which can lead to a better vocal performance for a song.

Many people associate the use of pitch correction with disposable pop music. Although there are many more factors in creating this pop music, pitch correction is used with every style of music. Even country and folk music, which has traditionally utilized vocalists with imperfections, such as Johnny Cash or Bob Dylan, has replaced their vocal stylings with a much more pop-oriented, perfect vocal sound.

The more that an engineer utilizes pitch correction in his or her work, the more he or she is able to hear the processing done on commercially released recordings. There is a noticeable sound to the pitch correction, but it is done so often that the general public has come to expect the perfectly pitched vocal.

As an engineer or producer, you should be well versed in being able to apply a subtle pitch correction to a vocal, as well as creating the "perfect" pop vocal. Different techniques with the various software plug-ins that are available allow for the different fluctuations to come through, while creating a more consistent vocal. It is all about using the right tool for the right job.

AUTO-TUNE

One of the most commonly used pitch-correction methods of the past ten years has been the use of the Auto-Tune plug-in, produced by Antares. There are two versions that are used in the studio. There is a stand-alone hardware version, as well as various plug-in incarnations of this technology. With the common usage of DAWs, the plug-in version has become the most common format utilized.

In order to get the most out of Auto-Tune, it is important to understand how pitch, as it relates to digital audio, functions. The pitch of a sound is determined by the period at which the waveform repeats itself. A monophonic, melodic instrument, such as the human voice, will create a periodic waveform (Figure 4.1).

The number of times that the periodic waveform repeats in a second determines the frequency of that pitch. Auto-Tune will analyze the pitch of the waveform in order to determine what pitch it is currently hearing, and then it will correct the pitch by stretching or compressing the periodic waveform and regenerating it to

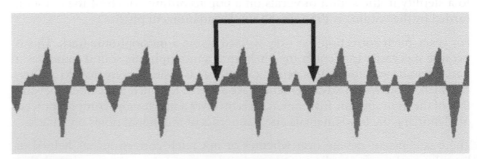

FIGURE 4.1
A periodic waveform with the arrows indicating a single period.

the desired correct pitch. For this reason, Auto-Tune will not work with any source material that contains polyphonic audio, such as a guitar chord or choir. When these waveforms are analyzed, there is no periodic waveform for Auto-Tune to detect what pitch it is hearing, and will therefore do nothing with the sound.

Auto-Tune's Automatic mode

There are two different modes of pitch correction with Auto-Tune: Automatic and Graphic. Automatic mode works well if you want to set the plug-in across the track with some general settings and let the plug-in work its magic.

When first opening the Auto-Tune plug-in, you will see that its default mode is set to "Automatic," with the "Chromatic Scale." The pitch-detection algorithms work best when selecting the Input Type that will match the source material. The choices are Soprano, Alto/Tenor, Low Male, and Instrument with the last choice of Bass being grayed out (Figure 4.2). This is because the Bass instance of the Auto-Tune plug-in is a specific algorithm designed to detect much lower pitches than are normally heard in the human voice and most monophonic instruments, such as a saxophone or cello.

When playing your track through the Auto-Tune plug-in, the piano keyboard (Figure 4.3) at the bottom of the screen will turn blue on the pitch that Auto-Tune is tracking from the source material. The bar above at the top will indicate the amount of pitch correcting that the Auto-Tune plug-in is performing. If the yellow indicator moves above the zero point, it means that the pitch is detected as being flat, and so Auto-Tune is raising the pitch by the amount of cents indicated.

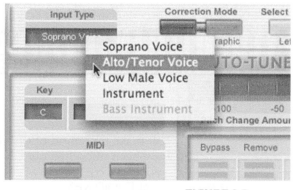

FIGURE 4.2
The voice selection parameter of Auto-Tune to select the appropriate algorithm for the material.

RETUNE SPEED

There are two main settings of the Auto-Tune plug-in that will have the greatest impact on the sound of the track. The first setting is the Retune Speed. This setting will determine how quickly the pitch is corrected, once detected, in milliseconds. An extreme setting of 0-ms will result in the famous "Cher sound" from her song "Believe." This extreme setting can also be heard in many other songs, such as "Only God Knows Why" by Kid Rock.

The more realistic settings begin at 20-ms for a solidly pitch-corrected sound and continue up above 100-ms for a more natural performance. Like many plug-in settings, Retune Speed can be automated in such a way that you can adjust the settings throughout the track to grab the pitch faster when necessary and to be more relaxed in its correction in other places.

If you are looking to minimize or remove a vibrato from a vocal performance, you can smooth out the pitch with a faster retune speed; however, there are other components to a vibrato such as formant and volume changes that will

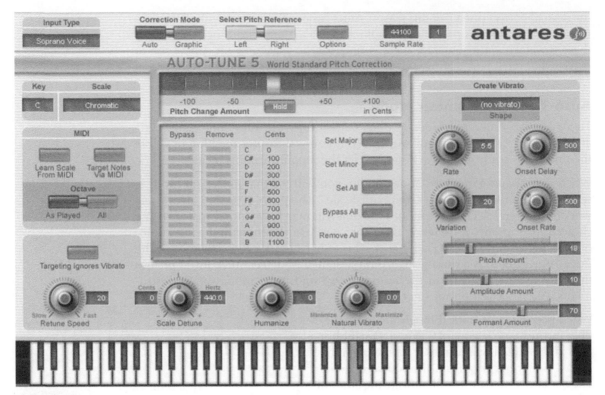

FIGURE 4.3
The keyboard in the Auto-Tune plug-in, indicating the detected pitch.

still make the sound of the vibrato present, even though the pitch fluctuations have been smoothed out.

SCALE AND KEY

The second-most important settings of Auto-Tune are the key and scale. This will determine which pitch Auto-Tune will correct the incoming pitch toward. If the vocal or instrument is fairly accurate, then the default setting of "Chromatic Scale" will probably work just fine. For instances where the "out-of-tune vocal" is actually closer to a different semitone, then Auto-Tune will correct the pitch to the nearest possible semitone, which could be a half-step above or below the desired pitch. In order to alleviate this, finding and setting the correct key and scale will eliminate the possibility of correcting the pitch to a note that is not even in the scale of the song.

The most common two scale settings for pop music are major and minor, but there is just about every scale imaginable in the scale settings. If you select "E minor" as the scale, you will see all of the tones that are not in the scale disappear as options (Figure 4.4). There are two options for creating a major or minor scale. The first option is to directly select the major or minor scale. The second, and most flexible, method is to select the correct key. Then select "Chromatic Scale" and press the "Set Minor" button that appears to the right of the notes (Figure 4.5). This will

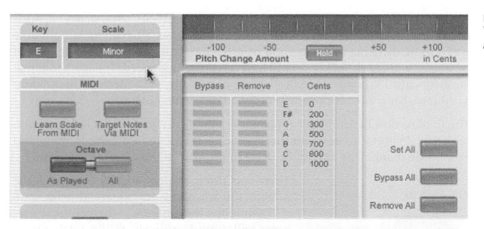

FIGURE 4.4
The scale selection of Auto-Tune set to E minor.

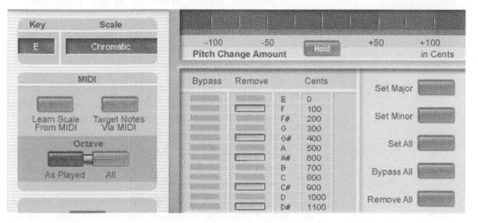

FIGURE 4.5
The key has been set to E minor with a Chromatic Scale by pressing "Set Minor."

remove all of the unused tones from the scale, but you will then have the option of unremoving a note. This could be used if the singer intends a single pitch that is not in the specified scale, such as adding an A sharp if the singer has a passing tone or the chord being played happens to not be diatonic to the selected key.

Another method of removing a note from the scale is by clicking on the piano key at the bottom of the plug-in. This will remove that single note and only in the octave it was played. This feature is quick and easy to use, as you can watch the keyboard below as Auto-Tune is tracking the pitch, and any erroneous pitch tracking is visible. These erroneous pitches can be quickly removed by pressing those particular keys on the piano.

ADDITIONAL FUNCTIONS IN AUTOMATIC MODE

There are three other knobs that control other features in Automatic mode. The first is Scale Detune. This will allow the user to make small adjustments up or down to the global pitch reference of the Auto-Tune plug-in. If the instruments that the vocalist is singing to are all in tune with each other, but are slightly below A440 as a standard pitch reference, then you can lower the Scale Detune control to match the pitch reference being used. This will prevent Auto-Tune from correcting the pitch to an incorrect pitch reference, such as correcting the vocal track to be sharper in pitch than the others. The next knob is the Humanize function. This will make the retune speed more dynamic to fit the vocal part. For longer notes, the retune speed will be longer, and the shorter notes will have a faster retune speed. The last additional knob is Natural Vibrato. This will allow the user to increase or decrease the amount of vibrato that is already present in the vocal or instrument track.

The final section in Automatic mode is the Vibrato section, which is designed to create a vibrato to a vocal part that has none. It will simulate the formant and amplitude changes, as well as the delay in the start of the vibrato (Figure 4.6).

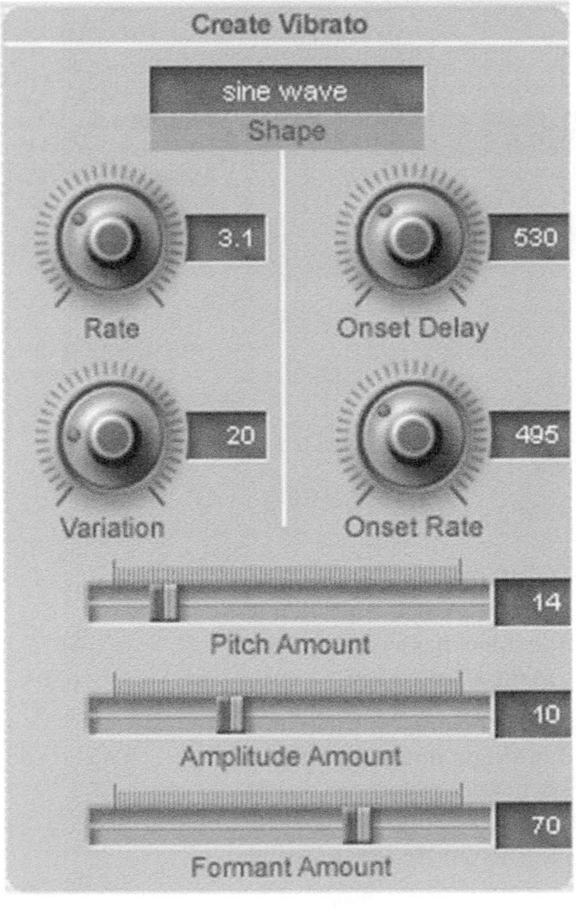

There are three choices for the shape of the vibrato: sine wave, saw, and square. The sine wave will sound the most natural, whereas the square wave will sound artificial as the pitch jumps up and down with little glissando between the modulations. The saw wave will sound artificial as well, because the vibrato moves up in pitch and then suddenly drops down.

The Onset Rate will determine the speed of the vibrato in hertz, or cycles per second. The Onset Delay will determine how long Auto-Tune will wait before engaging the added vibrato. The Onset Rate will then control how long from the end of the Onset Delay before the vibrato is at its maximum level. This will make the added vibrato seem more "human," as there is usually some time at which a singer will take to initiate the vibrato once singing the note. The Variation knob will further humanize the vibrato by adding random variation to the Onset Rate and Amplitude Amount controls.

The Pitch Amount will control the amount of depth to the vibrato; that is how far the pitch will go up and down in relation to the original sung pitch. The Amplitude Amount will create differences in volume, which is something that occurs subtly in a vocalist's natural vibrato. The Formant Amount controls the amount of resonance that is present in the original pitch.

FIGURE 4.6
The vibrato settings of Auto-Tune.

Graphic mode

Graphic mode in Auto-Tune is best suited for making specific and precise manual adjustments to the pitch correction. Within this mode, you can do more dramatic pitch correction than found in Automatic mode. What is important to note is that within this mode, you will need to process the audio by recording the pitch-corrected version onto the same track or a different track.

SETTING UP THE TRACK TO BE PROCESSED IN GRAPHIC MODE

Depending on the software used, it is best to record the processed audio track to a different track. To accomplish this, start by setting the volume in the mixer of your DAW to zero, and remove or deactivate any plug-ins and volume automation on the source track, with the exception of Auto-Tune. Once this is done, create a send to an unused bus from the source track. Create a new track for the processed vocal below the original track with the input being the destination bus set from the track being processed.

For example, if you have a track with a vocal needing Auto-Tune, create, in Pro Tools, "send the output to Bus 1," or any unused bus. "Create a New Track" and have the new track with its input from Bus 1. Mute this new track, so that when you are recording the processed audio, you do not hear both tracks playing at the same time. See Figure 4.7.

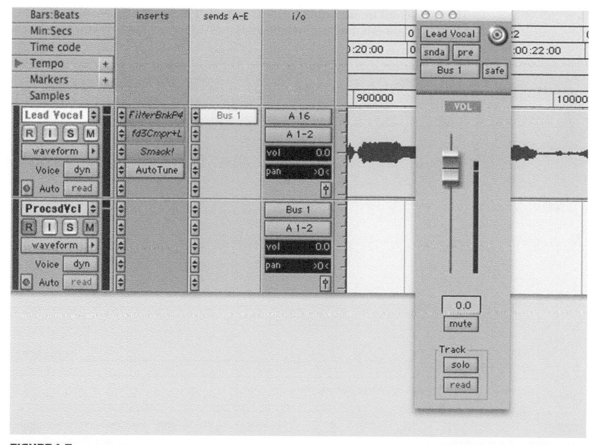

FIGURE 4.7
Setting up the track to record the processing of Auto-Tune.

THE TWO STEPS OF GRAPHIC MODE

There are two processes that you will need to do when setting up a track to be recorded with Graphic mode. The first is to Track Pitch, where you will select a portion of audio to be processed in your DAW. Then, the next step will be to Make Curve, which will lay a map down on the tracked audio to see what Auto-Tune will do to the audio. When first set up, the new curve laid down by Auto-Tune will correspond to the original tracked pitch, so there will be no change in pitch to the corrected audio track. From here you can manipulate the curve in a variety of ways to get Auto-Tune to correct the pitch as you would see fit (Figure 4.8).

When utilizing Graphic mode to make pitch corrections with Auto-Tune, it is best to work with single vocal phrases, rather than long passages of audio.

With the tracked audio in Graphic mode, you can move the entire curve around manually or redraw your own. With the Arrow tool, which is the default tool, once the pitch is tracked, you can drag the curve up or down to raise or lower the overall pitch. This is useful if you want to just raise a slightly flat vocal part overall.

The easiest tool to use would be the Line tool, which is the first tool on the left. From here you can draw segmented lines that will create a steady corrected pitch

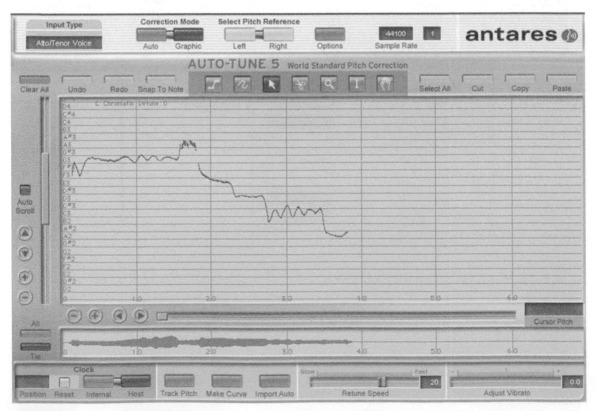

FIGURE 4.8
The pitch of a vocal tracked by Auto-Tune.

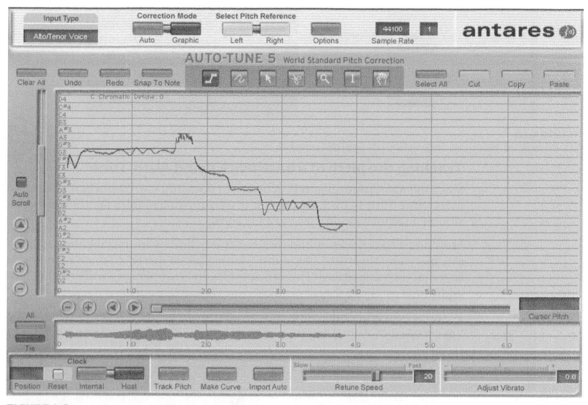

FIGURE 4.9
The corrected pitch has been drawn in with the Line tool.

where you have drawn them in (Figure 4.9). The Line tool will create a sequence of lines, segmented every time you click the mouse. To end the segmenting of lines, just double-click the mouse.

The next tool is the Curve tool. With this tool, you can draw in any corrective curve that you wish.

The Scissors tool will allow you to make cuts in a continuous curve that Auto-Tune has drawn on the screen. This is useful if you want to raise and lower different sections of the curve with the Arrow tool.

Once your Graphic mode settings are the way that you want them, then it is time to record the processed audio onto a different track (Figure 4.10). Simply place the newly created track into "Record," and record your selected segment. The audio region of the new track can then be placed onto the original track if you wish, or you can keep it in a different track, and then just mute the region of the original track.

Using Graphic mode in this way allows for a combination of both modes. You would be using Automatic mode for most of the track and then processing certain selections with Graphic mode, where more fine-tuning is necessary.

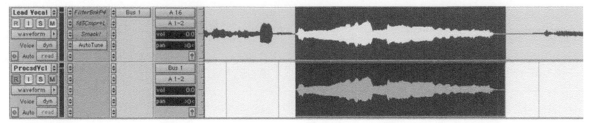

FIGURE 4.10
The pitch corrected with Graphic mode has been recorded onto a separate track.

Tracking the pitch of a bass track using Auto-Tune

Since the Auto-Tune plug-in is designed to track and correct pitch, it is possible to use Auto-Tune to track the pitch, if you are looking to track the chord progression of a section or find the key of the song.

Since most bass tracks in pop-rock songs rely heavily on playing the root, you can place the Auto-Tune plug-in in Graphic mode merely to analyze the pitch of the notes being played. This is helpful if you are looking to write or sequence additional parts to a song.

Without even processing the audio, you can see in Figure 4.11 the root notes that were tracked by Auto-Tune are C, Ab, Eb, and Bb. This is a great starting point in transcribing the chords to a song or figuring out the key.

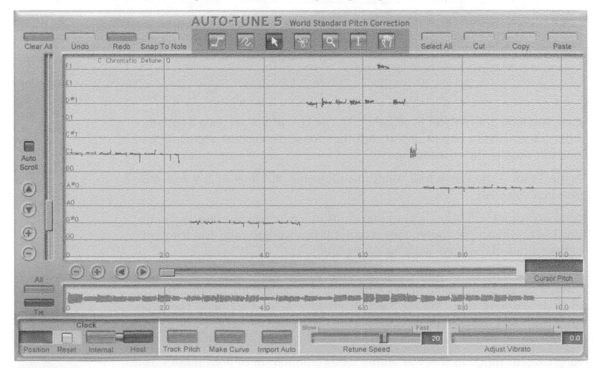

FIGURE 4.11
The bass notes tracked by Auto-Tune for transcription of the root notes.

MELODYNE

Melodyne is a powerful tool for pitch correction and more. It goes beyond pitch correction and allows you to adjust the volume, compress and expand note lengths, and create harmony vocals. Melodyne functions differently than any other real-time pitch-correction utility. Rather than correct the pitch in real time, Melodyne allows you to visually see a representation of the pitches and adjust them manually or automatically. This works well when you need to adjust just a few notes or the entire track. Melodyne is available in many commonly used plug-in formats, such as AU, RTAS, and VST.

Transferring the audio into Melodyne

In order for Melodyne to analyze the audio and process it, you will need to transfer the audio from a track on your DAW into a separate file on your hard drive. It is best to select the location on your hard drive that contains your session. This way, when backing up your session, any recorded files from Melodyne will be backed up with the rest of the session data.

Make sure any edits on the vocal track are completed before applying Melodyne. Because Melodyne will read transferred audio from a separate location outside of the DAW, it has no way of knowing if any edits have been done in the DAW once the track has been transferred. If edits need to be made on the track, you will need to retransfer the audio back into Melodyne for those sections.

An entire track can be recorded to Melodyne, but if it only needs to be applied to certain sections of the track, you can just transfer in those sections. When the track plays through the Melodyne plug-in, it will play back audio directly from the track until it reaches the point where the audio has been transferred into Melodyne, at which point it will switch to playing back the transferred audio automatically. After your DAW has played through the transferred audio, it will switch back to resuming the playback of the audio on the track in the DAW.

FIGURE 4.12
The Preferences setting of Melodyne.

Begin selecting Melodyne to be the first plug-in for the audio track. If the plug-in is instanced after a compressor or equalizer, and the processing on those tracks will be transferred, any adjustments you make will not be reflective on what you hear coming out of the Melodyne plug-in, due to the reading of the transferred audio. After the plug-in has been inserted, go to Settings and Preferences (Figure 4.12) and select the new location in your session for the recorded files for the "Temporary Recording" folder (Figure 4.13).

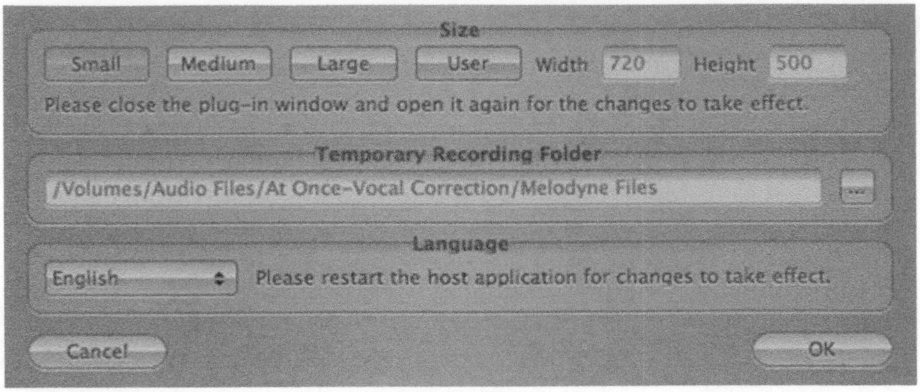

FIGURE 4.13
The selected recording
folder for the transferred
audio.

FIGURE 4.13
The selected recording
folder for the transferred
audio.

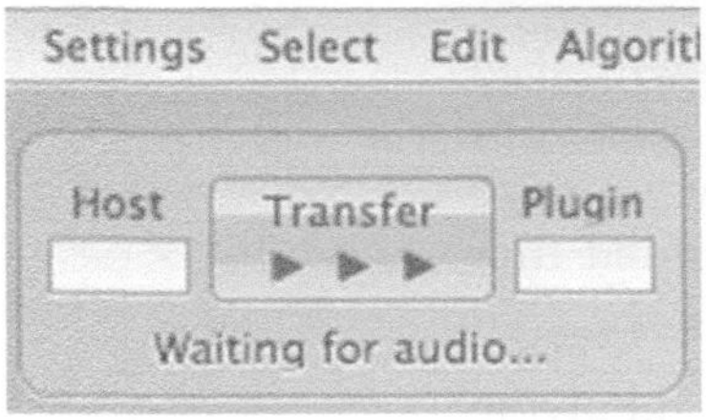

FIGURE 4.14
The "Transfer" button
pressed to record the
audio to the temporary
folder while playing back
the track.

Once the record path has been selected, you can begin to transfer the audio into Melodyne. This is accomplished by clicking the "Transfer" button, which you will see glowing red, and then play through the entire track or individual phrase that you wish to use with Melodyne. Playing back the audio will now transfer the audio files into the previously selected folder for processing.

Once the tracks have been transferred, Melodyne analyzes the audio for pitch, volume, and timing information and makes separations between the notes (Figure 4.15). Once these data have been gathered, you can begin to edit the pitch, timing, and volume of the individual notes.

Manually correcting pitch

To begin editing the pitch, you will need to select one of the pitch editing tools. There are three tools from which to select: Edit Pitch, Pitch Modulation, and Pitch Drift (Figure 4.16). Each of these tools has a specific pitch parameter that it adjusts, and we will delve into each of these tools.

Once the Edit Pitch tool has been selected, you can see Melodyne place gray blocks near the audio where Melodyne thinks that the corrected pitch should be moved to (Figure 4.17). Each pitch can be moved to the corrected location by double-clicking each note, which will then snap that note to the new location (Figure 4.18). If you make an error in any pitch correction, you can easily go and undo that move. Be sure that you undo the move from inside Melodyne's editor as opposed to your DAW's undo.

Because Melodyne averages the pitch of the notes when it makes its corrections, if there is some drift within the note itself, Melodyne may make the corrected pitch sharper or flatter during the sustained portion of the note (Figure 4.19). Trying to grab the pitch and manually adjust it will only snap the pitch to a neighboring note up or down. You can bypass Melodyne's note snapping by holding down the "Option" key while dragging the note up or down (Figure 4.20).

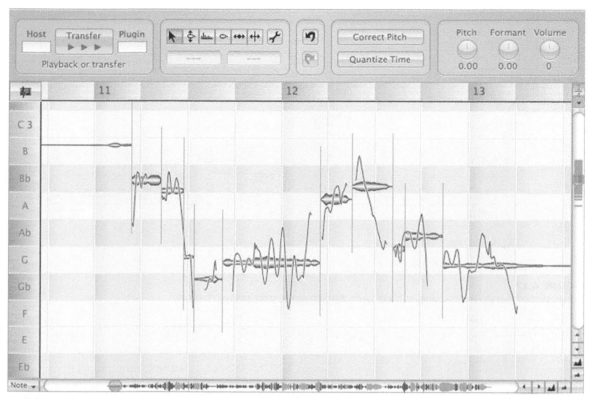

FIGURE 4.15
The tracked audio transferred and analyzed in Melodyne.

If at any time you wish to revert to the previous uncorrected pitch, you can highlight the notes with the Selector tool, go to the Edit window, go under Edit Pitch, and select "Reset All Pitch-Related Changes to Original." This can be done on individual notes or entire selected passages (Figure 4.21).

FIGURE 4.16
The three different tools for editing the pitch inside Melodyne.

Pitch drift and vibrato

When Melodyne corrects the pitch, it maintains the vibrato as well as any drifting between the notes. Essentially, the center of the pitch has been adjusted, while the rest of the singer's expression remains. The drifting between notes can be adjusted separately from the rest of the audio by selecting the Pitch Drift tool. Editing the pitch drift can increase or decrease the time it takes for the pitch to go from one note to the next. Since drifting between pitches is very natural for a singer to do (Figure 4.22), shortening the drift time can lead to unnatural-sounding results (Figure 4.23). The bigger the jump in pitch, the longer the drift should take for it to sound natural. For faster notes and shorter jumps in pitch, the drift can be shortened to whatever sounds best with the vocal.

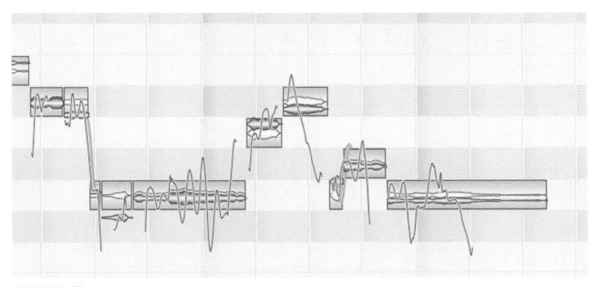

FIGURE 4.17
The detected pitch is shown as rectangular boxes, where the pitch can be snapped to automatically.

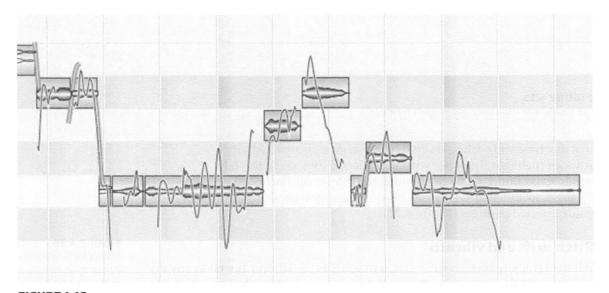

FIGURE 4.18
The transferred audio has been snapped to the corrected pitches.

Melodyne will not initially adjust the vibrato (Figure 4.24). It will just keep the pitch centered. If the singer's vibrato becomes too pronounced in specific sections, you can manipulate the vibrato with the Pitch Modulation tool. You can even extend or eliminate the vibrato completely. This is done by selecting the pitch and moving the Pitch Modulation tool either up, to extend the vibrato, or down, to minimize or eliminate it (Figure 4.25).

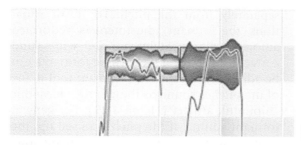

FIGURE 4.19
The corrected pitch, which is snapped to the detected pitch center, but is corrected sharp.

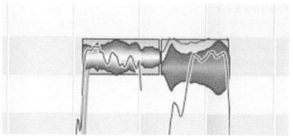

FIGURE 4.20
The corrected pitch has been manually adjusted so that it is no longer sharp.

FIGURE 4.21
Resetting the pitch parameters of all selected notes.

FIGURE 4.22
The natural, long drift associated with a larger jump in pitch.

Separating notes

Sometimes the singer may have long-held notes that change pitch. Melodyne may recognize these notes as being one single note, and so any adjustments made will affect all the sung pitches during the held note. You can use Melodyne's Note Separation tool to create a break in the long-held note so that you can manually adjust each pitch (Figures 4.26 and 4.27). Just double-click in the middle of the note where you want to add a note break (Figure 4.28). These separated notes can be edited in the same way as the other notes.

Editing formants

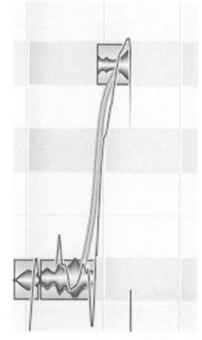

Formants are resonant frequencies that are generated by chambers inside the human head. These formants vary depending on the vowel sung. Generic pitch shifting, which just shifts the entire pitch up or down, will also shift the formants up and down. This will create an artificial or chipmunk-type sound when shifted up. When Melodyne adjusts the pitch, it keeps the formants intact and unshifted. This makes for a more natural-sounding pitch adjustment. Melodyne gives you the option of adjusting the formants

FIGURE 4.23
The pitch drift has
been corrected to 100
percent, making an
unnatural-sounding
instant jump.

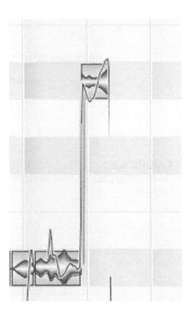

separately from the pitch. For most situations, the formant adjustment is performed automatically by Melodyne and will sound correct for the vocal part. These formants can be adjusted with the Edit Formant tool if any of the pitches sound as though they have this chipmunk artifact (Figure 4.29). For general formant editing, if the pitch is raised up, the formant will sound more accurate being lowered and vice versa for lowering the pitch and raising the formant.

By using the Edit Formant tool, you can raise or lower the formant just by dragging the mouse up or down. By adjusting these formants you can hear the resonances change, but the perceived pitch remains the same (Figure 4.30).

FIGURE 4.24
The natural vibrato visual
in the corrected pitch.

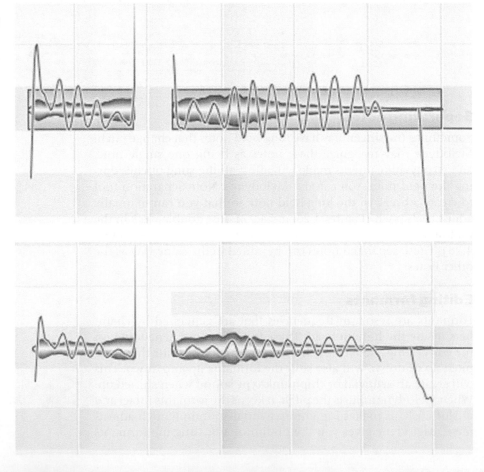

FIGURE 4.25
The natural vibrato has
been reduced, but not
eliminated with the Pitch
Modulation tool.

Automatic pitch correction

In addition to being able to manually adjust pitch with Melodyne, you also have the ability to correct the pitch of entire passages all at once. This can be accomplished by selecting the data you wish to automatically correct the

FIGURE 4.26
The Note Separation tool to break longer notes into individual notes.

pitch on, and then pressing the "Correct Pitch" button (Figure 4.31). You can select individual passages by using the Selector tool and dragging over the notes

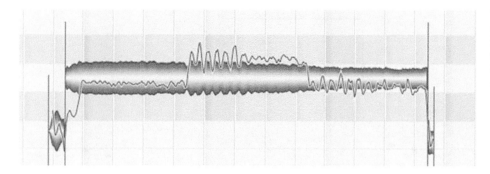

FIGURE 4.27
A long-held note, identified by Melodyne as being a long note, making both notes sound out of tune.

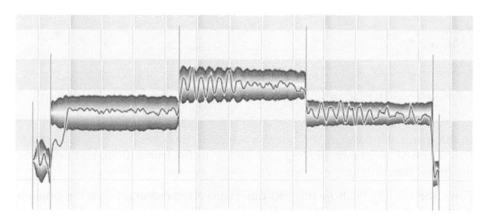

FIGURE 4.28
The held note has been broken into the individual notes, which can then be snapped to the appropriate pitch.

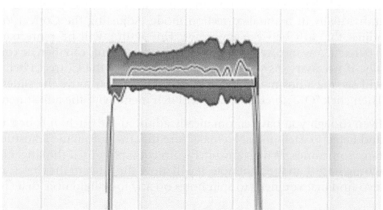

FIGURE 4.29
The formant that is visible when selecting the Edit Formant tool.

FIGURE 4.30
The formant has been lowered, while the corrected pitch remains the same.

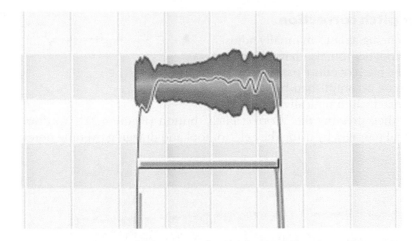

FIGURE 4.31
The automatic Correct Pitch menu, which will adjust all the selected audio.

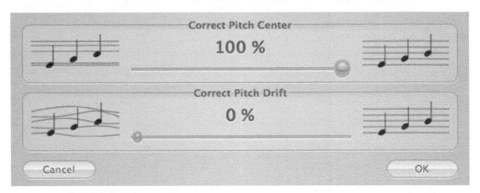

in that particular passage. You can also select all the notes that have been transferred into Melodyne by moving your mouse to the Select menu and dropping down to "Select All."

From the Correct Pitch menu, you can adjust the percentage that the original pitch will snap to the correct pitch. If the Correct Pitch Center is set to 100 percent, then Melodyne will adjust all the notes as if you had double-clicked on them individually in manual-correction mode. Adjusting the Correct Pitch Drift will adjust the drift between the notes. This setting will be more natural if left at 0 percent. Any incorrect automatic pitch adjustments can then be corrected manually, if necessary. As you adjust the sliders from the Correct Pitch window, you will see the adjustments made in real time as you move the slider up or down. Then click "OK" to exit the Correct Pitch menu with the adjustments made.

Even though you have automatically adjusted the pitch, it is best to go through and listen to each phrase to make sure that the automatic adjustments are accurate. Sometimes, as with any pitch-correction plug-in, if the singer is closer to the wrong note than the right note, it will move the pitch to the closest note. You may also find that you need to split notes on any long-held note that changes pitch.

Working with doubled vocals

If there are any tracks that are doubling the lead vocal, correct the pitch by using a separate instance of Melodyne in the same way that you would correct the lead vocal. Having the doubled-track pitch corrected in the same way as the

FIGURE 4.32
Melodyne's global pitch adjustment knobs.

lead vocal can almost make the tracks blend in too well together. The pitches may now be identical, with no slight fluctuations. If you are looking to create a thicker sound with the doubled vocal, you can fine-tune the pitch in Melodyne with the real-time correction knobs (Figure 4.32).

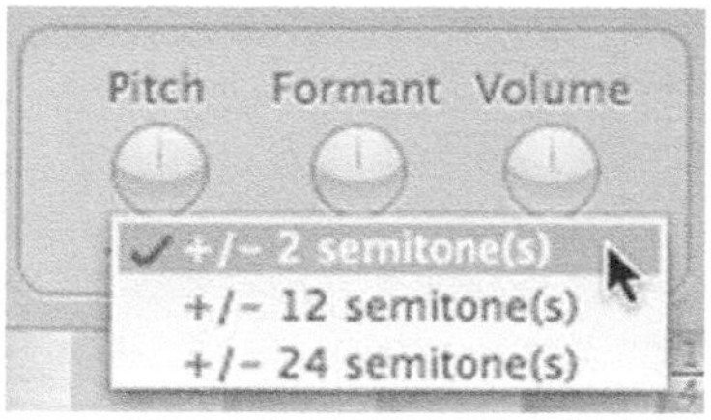

These correction knobs will apply their changes to the entire transferred audio. By right-clicking on the "Pitch" knob, you can adjust its range of resolution to being plus or minus two semitones (Figure 4.33). You can then detune the overall pitch by making fine adjustments with the Pitch knob. Detuning the track by a few cents will create a thicker sound than if two identically tuned vocal parts are playing simultaneously. See Figure 4.34.

FIGURE 4.33
Adjusting the range of the global pitch knob to two semitones.

FIGURE 4.34
The global pitch has been detuned by eight cents, creating a slightly detuned vocal track.

Working with scales

Just as with Auto-Tune, Melodyne gives you the ability to constrain the pitch correction to a particular scale. If all the notes on the track are diatonic to a specific scale, this can aid in the accuracy of the pitch correction as 5 out of the 13 notes of the chromatic scale are eliminated from the standard major or minor scale.

Melodyne is defaulted to snap the note to the nearest pitch and chromatic scale. If you know the key of the song, you can dial in the correct scale for the Melodyne plug-in. The scale can be adjusted by going to the Settings menu and selecting "Tone Scale." This gives you the option of major and minor scales, as

FIGURE 4.35
The menu selection for adjusting the scale in Melodyne.

well as several other exotic scales that could be used (Figures 4.35 and 4.36). There is also a selection for choosing the appropriate key (Figure 4.37). Lastly, there is the option to adjust the pitch reference. This is based on "A" being 440 Hz. This can be adjusted up or down depending on the pitch reference of the instruments, but typically pop recordings will all use A 440 as the standard pitch reference.

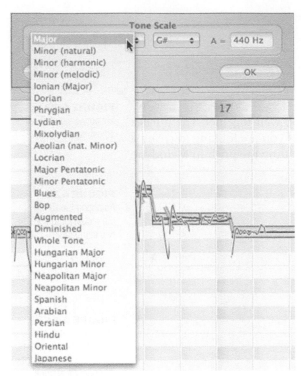

FIGURE 4.36
The vast scale selection in Melodyne.

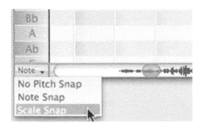

FIGURE 4.37
The selection to identify the grid where Melodyne will snap the notes.

You will see that the scale in the Edit window of Melodyne has not adjusted according to any changes made under the settings. This adjustment is made in the lower left corner where you change the default "Note Snap" to "Scale Snap." With Scale Snap enabled, all the notes still appear on the timeline; it is just that the selected notes will only snap to the nearest pitch in the selected scale. If there are certain notes that are not diatonic to the selected scale, you can manually adjust the pitch of these notes by holding down the "Option" key and dragging the note up or down to the appropriate pitch that may not fall in the designated major or minor scale. If you have selected an incorrect scale, Melodyne will show some of the targeted note boxes as being a semitone away from the actual correct pitch (Figure 4.38). You can click on the orange highlighted key on the left side of the Edit window to adjust the tonic of the scale. This can help you quickly make adjustments to find the correct scale if you have not been given the information.

Creating harmony vocals

Since Melodyne can adjust pitches in a natural way, you can utilize it to create your own harmony parts to the lead vocal. Suppose there is a line that might sound great with a harmony. You can use Melodyne to raise the pitch of that part up a third, or however you wish to create your harmony part.

To begin creating a harmony part, select the line you want to be harmonized and copy that region onto a separate, new track. Open up an instance of Melodyne across that new harmony track and transfer the audio in as before. Constraining the scale will allow you to manipulate the pitches up or down in the key of the song. You may not find that you can create harmonies everywhere. You may notice that the pitch change may sound artificial in places, as there is a substantial amount of pitch change being applied, but in certain key places in the mix you can certainly get away with highlighting different words or phrases by creating these artificial harmonies.

Bouncing the corrected tracks

Because Melodyne does not process the audio on the track, but actually reads in a separate file that has been transferred, the corrected track needs to be recorded onto a separate track with all of the pitch corrections. This is especially important after you have mixed the song and are looking to archive it.

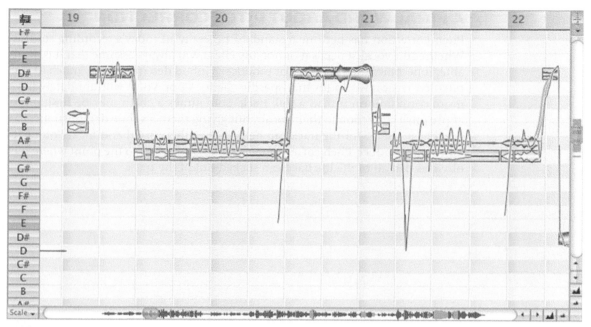

FIGURE 4.38
Melodyne identifying wrong pitches due to an incorrect scale setting.

In order to record the processed vocal track, make sure that any plug-ins, with the exception of Melodyne, are deactivated. Make sure that the audio output of the fader is set to zero; otherwise, there will be a gain difference in the bounced track. Send the lead vocal track through a single auxiliary bus instead of the stereo mix.

Create a new track and have the input set to be from the bus of the output of the corrected track. Once the corrected track has been recorded, you do not need to eliminate the track with Melodyne across it; you can merely mute it in case you need to apply further pitch correction in the future. See Figure 4.39.

FIGURE 4.39
Recording the pitch-corrected audio onto a different track.

A FINAL WORD ABOUT PITCH CORRECTION

Pitch correction has become a standard necessity in today's recording studios. Whether the vocalist needs it or not, the client will always assume that it is available. It becomes the engineer's or producer's job to determine how much of this pitch correction works for that specific client. A pop vocal will ultimately have more correction on it than a folk track. Sometimes a client may be hesitant to apply pitch correction to their vocals, but giving them a subtle demonstration as to what it can do for the track can help change their mind and make for a better overall product. Pitch correction *should not* be applied to the point where it is obvious that all of the life has been sucked out of the track.

CHAPTER 5
Emulated Effects

Since much of recording has moved from the analog studio, with all outboard analog equipment, into the world of digital audio workstations (DAWs), companies have been working on moving the outboard analog equipment into these DAWs. Emulated effects are not new to digital technology; for example, Tech 21 has been making the SansAmp, which contains all analog circuitry to emulate guitar amplifiers, for years. However, today there are many pieces of software that are designed to emulate their analog counterparts. These include equalizers, compressors, guitar rigs, keyboards, and reverb units. Sometimes these software plug-ins are general emulations, and other times they are licensed emulations from the original manufacturer.

These emulated effects can be more processor intensive than a general equalizer found in your DAW. They can also vary significantly from the actual sound of their analog counterparts. As technology has progressed, so has the quality of these emulated effects. While they may not accurately represent their analog counterparts, they can be very close.

GUITAR CABINET EMULATORS
One of the most powerful emulations available for use in a DAW is the emulation of a guitar cabinet. Guitar amplifier emulation is not new to DAWs or even in the analog domain. Before the proliferation of computer-based recording, there were several analog guitar cabinet simulators, and the quality of these varied greatly. This holds true for their digital counterparts, although the quality has dramatically improved over the years.

Guitar cabinet emulators can function well for both clean and distorted sounds. As mentioned previously, when you are recording, it can be a good idea to capture a direct sound from the guitar as well as a microphone on the guitar

FIGURE 5.1
Line 6's Amp Farm with licensed names of guitar amplifier models.

cabinet. Since guitar rigs can be very complicated with different cabinets, pedals, and other effects, guitar cabinet emulators can vary in their capabilities.

Some guitar cabinet emulators are designed to emulate a specific type of guitar amplifier. Some manufacturers may have actually licensed the names of the guitar amplifiers that they are emulating (Figure 5.1), while others have more generic names. Since a basic guitar rig consists of an amplifier and speaker cabinet, these two have the ability to be changed independent of one another.

Using a plug-in such as Line 6's Amp Farm or Digidesign's Eleven, there are knobs on the emulator to mimic those of their real-life counterparts. This can include any tremolo or vibrato added by the amplifier. Using an emulation such as this can accurately represent the feel of a guitar amplifier at least in terms of the way you would dial in the settings.

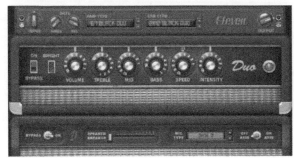

FIGURE 5.2
Digidesign's Eleven guitar cabinet simulator.

Some guitar cabinet simulators can opt to output the signal as mono or stereo. With a traditional input such as a guitar direct input (DI), the input is, of course, mono. Having a guitar cabinet emulator's output set to stereo will not necessarily lead to a dramatic width in the stereo image. However, there are stereo effects that can be added through the use of these emulators. These include phasing, flanging, and chorusing. If you are going to be using a separate plug-in for these types of effects, then you can merely select a mono output. If the plug-in that you are using creates these stereo effects, then select a stereo output.

One common feature among guitar cabinet emulators is a noise gate. This is designed to eliminate any hum that goes through your direct input. Sometimes a DI box or guitar output can be susceptible to noise. This noise can create a constant stream of distorted noise coming out of your guitar cabinet simulator.

Some guitar cabinet simulators also allow you to adjust the type of microphone used on a cabinet, as well as it's positioning. This is a good opportunity to hear how an emulated version of this microphone will sound on a guitar cabinet (Figure 5.3). Many of these different microphone emulations are close to how their real-life counterparts would sound on a guitar cabinet.

You can oftentimes select whether to put a microphone "on axis" or "off axis." Having a microphone placed "on axis" means that it is perpendicular to the guitar cabinet. This will have a brighter sound as opposed to placing the microphone "off axis," which will have the microphone angled in toward the speaker cone. Many guitar cabinet simulators will also have the ability to place the guitar microphone in a room. This will give the guitar sound more space, but it will ultimately not be as present in the mix. See Figure 5.4 for an example.

Guitar cabinet simulators can also contain effects that can be obtained through the use of other plug-ins. These include compression and equalization. If the equalization is part of the sound of a guitar cabinet, then use that equalizer as it may be designed for that particular guitar tone. A compressor, on the other hand, can easily be emulated with a separate plug-in. This may give you more options for the compression, as well as having a variety of different choices sonically.

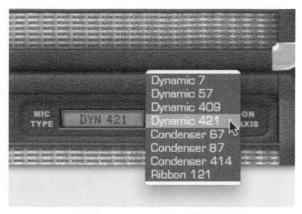

FIGURE 5.3
The different emulations of the microphones placed on the guitar cabinet.

Guitar cabinet emulators can also have a reverb that can be used. This reverb can easily be added through a different reverb plug-in. The difference is that there may be some reverbs that are tailored specifically to guitar cabinets. Any reverb that is built in a guitar cabinet is a spring reverb, and this can be considered part of the classic guitar sound. Many of the standard reverb plug-ins on the market may not have a spring reverb setting. Utilizing the spring reverb emulation from a guitar cabinet simulator can help achieve the desired classic guitar amplifier sound.

Guitar cabinet emulators can vary in their complexity. Some of them will even provide the option of adding chorusing, tremolo, and wah effects. These can be part of the plug-in itself or an additional option to add as a separate plug-in. McDSP's Chrome Tone comes with these additional options (Figure 5.5). You also have the ability to open up every available effect of this plug-in by selecting the Chrome Stack. The Chrome Stack gives the option of using any of Chrome Tone's modules and turning them on and off as necessary.

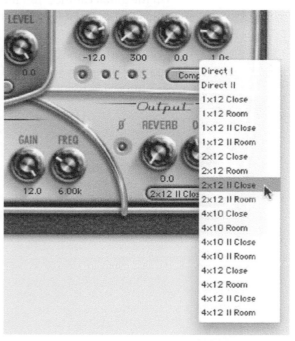

Doubling a guitar using a guitar cabinet emulator

If you plan ahead when you are tracking the guitars, you can choose to capture the direct input in addition to the microphone on an actual guitar cabinet. You can then run the direct input sound through the guitar cabinet emulator. You can do this for one of two reasons. The first is to create a stereo sound using two different guitar sounds.

FIGURE 5.4
McDSP's Chrome Tone, demonstrating different general cabinet emulations and microphone positions.

FIGURE 5.5
McDSP's Chrome Tone
guitar amplifier emulator.

Start by placing the direct input into the guitar cabinet simulator; you can then dial in a similar setting to the guitar player's amplifier. Hard panning both of these tracks can give you a wider sound than if you just place two microphones on the guitar cabinet. You may need to do some creative equalization on both of the tracks in order to get them to sound close enough so that they are believable as a stereo instrument.

You can also choose to layer two different guitar sounds with the same performance. This gives you more freedom when dialing in the simulated guitar tone as opposed to trying to create a stereo sound (Figure 5.6). You can choose a com-

FIGURE 5.6
A guitar cabinet emulator
placed across a direct
input track, doubling
the microphone on the
actual amplifier.

pletely different sound for the simulated guitar cabinet. For instance, you can use a twangy Fender Twin sound on top of a warmer and rounder clean tone to get the guitar to cut through the mix more.

Using guitar cabinet emulators on vocals

Since guitar cabinet emulators give you the ability to place a plug-in across any track, it has become commonplace to use a guitar cabinet emulator on other instruments to create a specific distortion effect. You can run into a whole separate set of issues when running a vocal track through a guitar cabinet simulator. The added gain that a guitar amplifier applies to the input signal is what creates the distortion to begin with; it can make even the quietest sounds on that track very audible once you place the track through a guitar cabinet emulator.

To prepare a nonguitar track to run through a guitar cabinet emulator, you can adjust the noise gate of the plug-in itself, or give yourself more flexibility by using a separate gate prior to the guitar cabinet simulator. Using a gate will let you adjust the attack, hold, and release times in order to make enough of the vocal track come through distorted, while not having every breath that the singer takes as loud and distorted as the actual words.

If you are planning on having a guitar cabinet emulator across part of the vocal track, it is helpful to have the distorted part on a separate track rather than automate the distortion to turn on and off. Since running distortion across any track can drastically affect the gain, you may find yourself doing more work to adjust the levels as well as the settings on any other plug-ins to accommodate the change in tone added by the guitar cabinet emulator.

Having experience running guitar cabinet emulators across the vocal track can give you more tools for production while making a record. You can plan in advance to have the lead vocal doubled with a separate distorted sound. This can create an interesting effect in the mix. The guitar cabinet emulator can also be placed in parallel with a clean vocal track. This will create a distorted, yet intelligible, vocal sound.

Guitar cabinet emulators on bass

Since guitar amplifiers and bass amplifiers sound different from each other, and you are looking to emulate the bass cabinet with a guitar cabinet simulator, you may find that it is difficult to do. You certainly can get some of the grit from the amplifier by using a guitar cabinet emulator, but it will not be the same as an actual bass amplifier. You can, however, create interesting distorted bass tones by using these plug-ins.

Guitar cabinet emulators on drums

Placing distortion across the drums using a guitar cabinet emulator can produce a trashy effect on the drums. Since drums can span several tracks as opposed to one track for a guitar or bass, it becomes tricky to apply distortion on all the

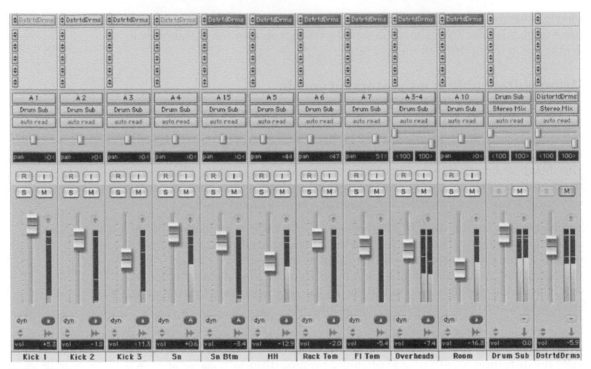

FIGURE 5.7
The routing of drums through a distortion on a separate bus.

tracks. The best way to have distortion across the drum tracks is to set the output of each track to go out a separate bus (Figure 5.7). This also gives you the added benefit of being able to control the gain of the entire drum kit up and down with that single master drum output.

Once the drums are dialed in the way you want it to sound, you can create a separate output to the drums by creating an auxiliary send across each of the drum tracks. Set each of the auxiliary sends to zero and make sure that they are all set to be postfader. This will ensure that the level of the auxiliary send that will be going to the distortion is the same as the drum subgroup. Create an auxiliary master track, which will serve as the distorted drum track. Now you automate back and forth between these two tracks and create a distorted effect, which can be used in different places in the song.

Once you have all the drums running into the distortion, you may find that the cymbals, hi-hat, and perhaps other drums may be making the distortion too piercing. This is due to the nature of the way distortion works. Since we have the tracks that are through the distortion coming from an auxiliary send, you can adjust the balance or even mute certain tracks to remove these sounds from the distortion. Again, there is an issue of the perceived loudness of distorted drums versus undistorted drums. This can be dealt with by adjusting the levels between the drum subgroup and distorted drums auxiliary master.

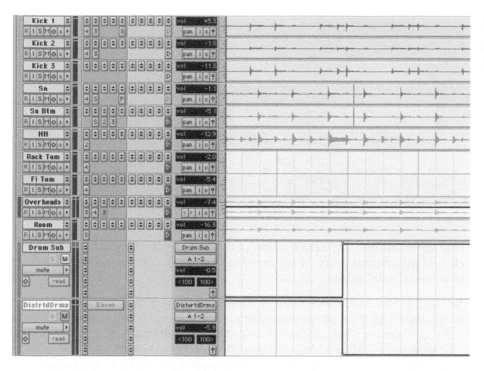

FIGURE 5.8
The automation of the drum distortion, going from clean to distorted.

Once you have the balance set between these two faders you can then automate the muting of each fader so one turns on when the other one turns off (Figure 5.8). This may not happen on the downbeat in your grid, so you may need to nudge the mutes earlier or later in the timeline.

EMULATED COMPRESSORS

Emulated compressors were some of the first plug-ins to emulate their hardware counterparts. This was due to engineers making the transition from working in an analog studio to working in a DAW. There was always a need for them to adapt something they were using in the analog domain to the digital domain. So much of a mix's sound is the type of equipment that is used for compression and equalization.

There are many more classic compressors made in the 1960s and 1970s that have a very high value in the studio today. Being able to have these compressors in a plug-in format rather than spending the thousands of dollars it would take to own one of these pieces of gear is one of the best money-saving benefits of a DAW. You also have the ability to use that plug-in on as many tracks as your DAW will allow. If you own one of these classic pieces of hardware in your studio, you can only put them across one track. See Figure 5.9 for an example.

The problem with modeling vintage analog equipment is that seldom are there two pieces of the same model that sound exactly alike today. In addition,

FIGURE 5.9
Bomb Factory's Fairchild
660 compressor
emulator.

there are oftentimes different versions of the same piece of equipment that were released in different years. These may look identical, but certain versions have the more "classic" sound.

The quality and accuracy of many of these plug-ins is a matter of debate. Some manufacturers go through the process of modeling every signal component inside of the compressor. This can be a time-consuming process, but it will yield the most accurate results. In addition to modeling the electronics inside a compressor, these plug-in companies will also faithfully reproduce the look and behavior of the compressor. Some manufacturers will release their own version of their compressors in plug-in format, such as Universal Audio. Others, like API, will license the use and name of their equipment to a plug-in manufacturer, such as Waves.

A good plug-in compressor emulation will emulate the way that a VU meter moves as well as making sure that the controls are familiar. In addition, the anomalies of the different compressors are modeled. For example, an emulation of the UREI 1176 will model what has been deemed the "British mode" (Figure 5.10). This is an interesting setting for the compressor where all four of the ratio buttons are pushed simultaneously. Once these buttons are pressed, the needle of the VU meter pegs all the way to the right. This has become such a unique effect that even analog compressors, such as the Empirical Labs' Distressor, can emulate this mode.

FIGURE 5.10
Bomb Factory's BF76
modeled compressor in
the "British mode" with
all four ratio buttons
pressed.

FIGURE 5.11
McDSP's CB4
demonstrating the
emulation of many
different compression
types.

Some compressors may not have the licensed emulations or possess the ability to emulate many different compressors with the same plug-in. An example of this would be McDSP's Compressor Bank CB4. This plug-in models the different mechanisms used in a variety of compressors including optical, solid-state, and tube styles of compression (Figure 5.11). Since there are many different mechanisms for compression, this allows you to choose the sound that works best for the track.

EMULATED EQUALIZERS

Equalizers are modeled in the same fashion as compressors. There are certain equalizers that are much sought-after pieces of gear in the studio. Just because equalizers are designed to mainly alter the frequencies across a track does not mean that there is not a difference in sound from one equalizer to another. Equalizers will have their own response as it relates to the Q and boost/cut. The electronics inside an analog equalizer will also play a role in the equalizer's sound. Some equalizers, just like compressors, are either tube or solid state. A modeled equalizer will reproduce the sonic characteristics of these components. See Figure 5.12 for an example.

MODELED DELAYS

There are different types of modeled delays. Since there are different methods of achieving delay in the analog domain, and each one has its own unique sound,

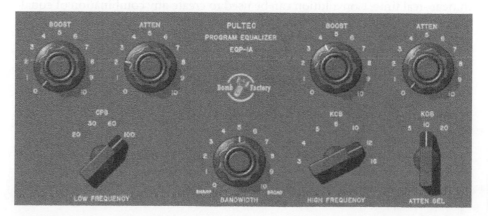

FIGURE 5.12
The Pultec EQP-1A
program equalizer
manufactured by Bomb
Factory.

these different methods of creating a delay are modeled in the various plug-ins. One of the first technologies for creating an analog delay was the tape machine. These tape delays record the audio on the record head and playback off the playback head. The difference in time between when the audio hits the record head and when it is played back from the playback head is a difference between the distance as well as the speed of the tape.

The other main form of analog delay that is modeled by plug-ins is the solid-state, bucket-brigade style of delay. This delay is created by bucket-brigade chips that pass the signal from one to the other, with each pass taking a short amount of time. The more of these chips in the device, the longer the delay will be. This type of delay creates a low-pass filtering effect on the delayed signal. This delay is emulated by Bomb Factory's Moogerfooger plug-in.

Delay emulation is not as common as equalizer or compression emulations, but there are some companies that create plug-ins that will emulate tape delays as well as bucket-brigade delays.

CONVOLUTION SIGNAL PROCESSING

Convolution, in DAWs, is one of the most powerful tools for emulating acoustical sounds. It can take the actual sounds of an acoustic space such as a cathedral or gymnasium, and allow you to reproduce them accurately. This gives you the power of bringing these acoustic spaces into your DAW. Convolution can also reproduce sounds from outboard equipment. You can sample an equalizer, preamp, delay, or reverb and place that sound across any track inside your DAW. Convolution processing uses an impulse response (IR), which can be derived from an acoustic space or piece of outboard equipment. There are specific software applications that are designed to create these impulse responses by deconvoluting the output of a piece of equipment or the recorded sound of a space.

Convolution is a fundamental process of digital signal processing. Convolution processing of audio is very processor intensive, and the results need to be output in near–real time. Convolution can be used to create any combination of equalization filtering, delay, and reverb. What convolution does essentially is take an input signal and process it through an impulse response to create a convoluted output.

There are limitations to convolution. With convolution processing in audio, you cannot adjust the pitch or create harmonic content that is not present in the original audio. It works best on capturing and reproducing the delay and frequency response of different equipment and environments. There are numerous web sites where impulse responses can be found from various pieces of equipment, spaces, and other odds and ends such as telephones and tiny speakers. There are many different convolution plug-ins available across any platform. Since convolution is a very CPU-intensive process, it is only in recent years that convolution plug-ins have been available for DAWs.

Convolution, simplified, is a multiplication of signals; the input signal is multiplied by the time and frequency content of the impulse response. The output is the frequency spectrum over time of the impulse response multiplied against the input. If the impulse response contains only low-frequency information and the input signal contains high-frequency information, with no overlap of the frequency content, then there will be no audible sound from the output. See Figure 5.13.

FIGURE 5.13
A frequency "waterfall" demonstrating the time and frequency composition of an impulse response.

Acoustic spaces through convolution

Using a convolution processor to create reverb is the most common usage for a convolution plug-in. Oftentimes convolution plug-ins are marketed as a reverb convolution processor. This does not mean that these processors are only useful for creating reverbs; however, most of the added features of convolution plug-ins are designed to tailor the sound of a reverb in much the same way you would modify settings with a stand-alone digital reverb unit. You have the ability to equalize the convoluted sound, adjust the decay time and early reflections, and adjust many other reverb settings you would find in a conventional reverb plug-in.

Since convolution processing is very processor intensive, you may find that you may not be able to add many instances of the plug-in before running out of horsepower. The longer the impulse response, the more processing the convolution plug-in will use up.

Many of the impulse responses used with convolution signal processing are created from actual sampled acoustic spaces. There can be many variations as to where the microphones were placed, what types of microphones were used, and what the distance was between microphones. Convolution plug-ins come with their own set of impulse responses, but the nice thing about using a convolution processor for your reverb is that you can always add new sounds, as opposed to using a stand-alone reverb unit that may be limited. In addition to the sounds distributed with the plug-in, most convolution processors have the ability to import other impulse responses that are in the form of an audio file.

Since the impulse responses are merely audio files, convolution reverbs make excellent choices for surround sound. There are many impulse responses that are available in 5.1 channels. This can be accomplished by sampling the space with four, five, or six microphones. See Figure 5.14.

Filtering using convolution

Because impulse responses can map the frequency resonances of a particular piece of equipment, you can easily use a convolution processor to mimic the sound of audio going through a telephone, speaker, or any other device that filters the sound (Figure 5.15). Utilizing impulse responses of these different

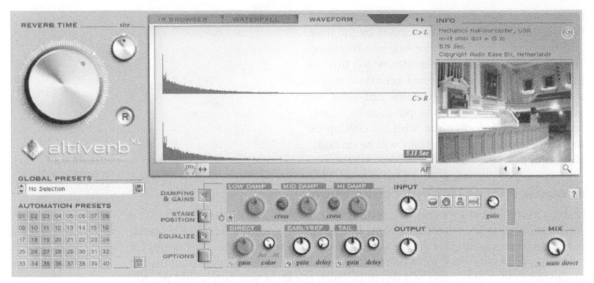

FIGURE 5.14
The Altiverb convolution processor displaying the impulse response as well as the various adjustable parameters.

devices is the best way to mimic their sound. There are many times when a client may want to have his or her vocals sound as if they were coming from a radio. You can accomplish this with some high-pass and low-pass filtering, but the end result does not have the same resonances as it would if you had actually used an impulse response created from a radio.

Even if you are not looking to sample your own devices and spaces, there are plenty of resources on the Internet where users will post their own impulse responses. You can quickly sort through these with your convolution plug-in to determine which ones are good and which ones are not.

Convolving impulse responses of external equipment

FIGURE 5.15
An impulse response made from a telephone, equalizing the audio to sound like it is coming from a telephone.

In much the same way that you can use convolution to reproduce the sound of an acoustic space, it can be used to reproduce the sound of any outboard reverb or delay. There are many classic sounds such as a plate or spring reverb that have a unique characteristic. Most nonconvolution reverb plug-ins will be able to emulate a plate reverb, but with a sampled plate reverb you can achieve more realistic results. Most home studios do not have the space for a plate reverb as many of them take up several feet worth of space.

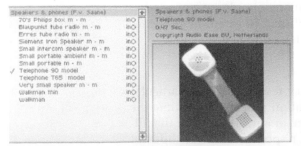

Many different reverb processors have been removed from studios so they can be sampled and used with a convolution reverb plug-in. This also gives the flexibility of creating more than one instance of that same reverb.

Creating your own impulse response

Most of the higher-end convolution plug-ins will come with a utility to sample your own acoustic space or outboard equipment. The process for sampling your own space begins with creating a source for the impulse response. The recorded response of the room or outboard equipment is then run through a deconvolution utility, which creates the impulse response based on the source material.

There are a couple different methods of capturing the sound of an acoustic space. The first of which is to use a starter pistol. This creates a burst of noise that reverberates throughout the room and is captured in the microphones. The other method is to use a generated swept sine wave. The sine wave is then played back off a CD into a speaker placed in a room. The microphones capture the sound of the echoed frequencies, and that recording can then be deconvolved into an impulse response.

GENERATING YOUR OWN SWEPT SINE WAVE

Convolution programs that come with their own deconvolution utility will also come with a utility to create a swept sine wave (Figure 5.16). There are oftentimes different formats for swept sine waves. If you are looking to create a high-resolution sample that goes above 44.1 kHz, then you need to select a sine wave that will go above 44.1 kHz. The sound file can be played back from a DAW or CD and then recorded into a portable recorder or back into the DAW. The more accurate your playback system is, the more accurate the final impulse response will reflect the sound of the room. The reason for this is that you will also be capturing the frequency response of the playback system. Having a playback system with small speakers that will not produce low frequencies means that there will be no low frequencies played back in the room from the swept sine wave, and so there will be no low frequencies to deconvolve.

RECORDING THE SINE WAVE

Recording the output of a swept sine wave does not create the impulse response directly. The recorded output of that swept sine wave, which can be a stereo or multichannel file for surround sound, is then sent through a deconvolution utility. The convolution utility is then set to identify the input source, whether it be a swept sine wave or a starter pistol. You may also have the ability to apply an equalization curve to the input file. This is to compensate for any playback issues of the original sine wave. Perhaps you used a playback system that did not have a flat-frequency response.

The swept sine wave can be played through a playback system in the room where you are looking to capture an impulse response. The microphones should be placed in a location away from the sound source, where you will be capturing the best

FIGURE 5.16
Altiverb's swept sine wave generator to create an audio file to sample a space.

Altiverb Sweep Generator

30 second sweep + 7 seconds silence

File Type
- for Audio CD Burning
- 44.1 kHz, 24 bits Mono SD2 file
- 48 kHz, 24 bits Mono SD2 file
- 88.2 kHz, 24 bits Mono SD2 file
- 96 kHz, 24 bits Mono SD2 file

Level
- 0 dBfs
- −12 dBfs
- −24 dBfs

Quit   Sweep

FIGURE 5.17
The Altiverb IR Pre-Processor, which derives the impulse response from the recorded space.

ambience. The polar pattern of the microphones used should be set up to capture the most ambience possible. This can include setting the microphones to an omni polar pattern, or directing a cardioid polar pattern toward the walls that would have the most reflections. The sweep can be captured from various locations in the room to create multiple impulse responses with varying audible distances.

When recording a swept sine wave, make sure that you are recording longer than the actual duration of the sound file, as you are looking to capture the complete decay of the audio in the room. The deconvolution software will be able to truncate the sound into the appropriate length.

Recording a swept sine wave, through outboard equipment, is easier than capturing an acoustic space. You can do all of the capturing through a DAW. The swept sine wave is placed in an audio track and then output into the piece of outboard gear. The output of the outboard gear is then recorded into a separate audio track in the DAW.

DECONVOLVING THE SAMPLED SPACE

FIGURE 5.18
The swept sine wave, above the recorded result of a sampled reverb effects processor.

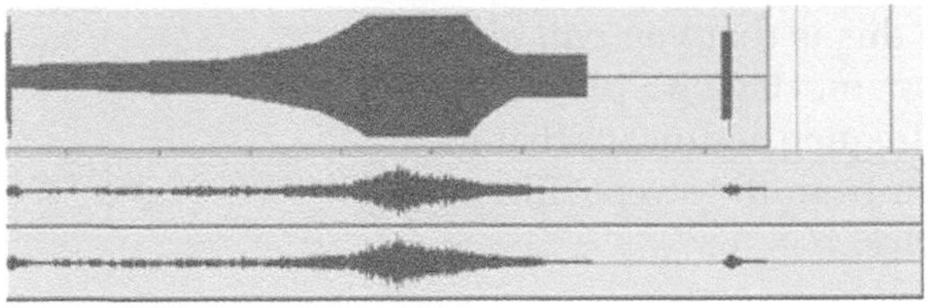

The end result of recording the swept sine wave is the creation of the impulse response. This can be accomplished by importing the recorded sound into a deconvolution signal processor (Figure 5.17).

You can compare the sound and quality of deconvolution by deconvolving a standard reverb plug-in. This gives the ability to run a single source through both the convoluted signal processor and the original reverb source. This will give you an idea of how close convolution comes to capturing the sound of an authentic acoustic space. Of course, you can sample different presets for your other plug-in reverbs, however, you will find that a convolution plug-in will take up more CPU power than the original reverb plug-in. See Figures 5.18 and 5.19.

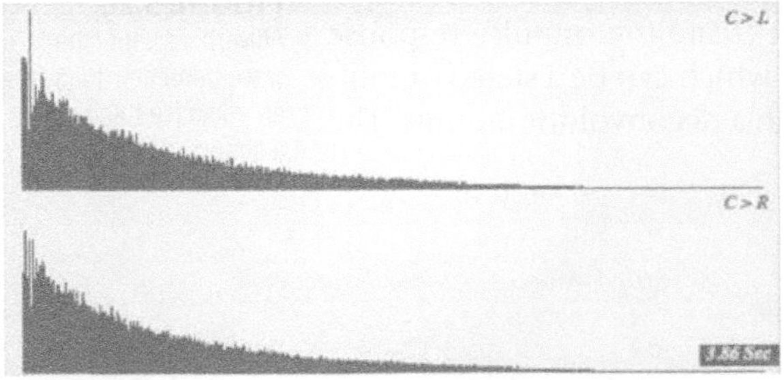

A FINAL WORD ABOUT USING EMULATED EFFECTS

FIGURE 5.19
The final impulse response, derived from the recorded sine wave run through the preprocessor.

Despite the quality of emulated effects today, these software plug-ins are around because of the innovation and quality of their analog counterparts. There still is something to be said about being able to place a high-quality compressor or equalizer across the track, even if it is coming from a DAW. This is what keeps some of the larger studios in business. As technology progresses, the outboard racks of these larger studios will continue to shrink, but there will always be a market for high-quality analog equipment for the discerning engineer.

CHAPTER 6

Adding MIDI Tracks to Recordings

CREATING PERCUSSIVE AND RHYTHMIC LOOPS

Incorporating MIDI tracks into a recording session can be a professional addition to many recorded projects. If done right, it can help a recording sound like a record as opposed to a demo. Even if the band does not have a "keyboard player," MIDI tracks can be added to supplement the song. You can reinforce chords, harmonically, by adding a low-level keyboard pad. Adding drum loops in a few sections can add an interesting element to the music.

Adding drum and percussion loops to a song

There are many software plug-ins that can add drum and percussion loops to a recording. These loops can be sequenced in time to the tempo map of the song (refer to Chapter 1 to read how to create a tempo map for a song by using Beat Detective). When selecting where to add these percussive loops, select a section that is not as busy rhythmically and could use an additional element. When placing a drum loop across a track that already contains drums, an added loop should not sound like the drum track; otherwise, it will sound as if there are two drummers playing. Instead, select an element that does not sound like real drums, but will add a complementary percussive sound for the specific section. Loops that contain a lot of low-frequency information can clutter up the bottom end of the frequency range. Percussive elements that are mostly high-frequency content will stand out in the mix more.

Rhythmic sequences using Stylus RMX

Stylus RMX is a software-based instrument, made by Spectrasonics, specifically designed for creating percussive loops. It has many different loops to choose from, as well as several ways to modify these elements (Figure 6.1). Many of the grooves that it comes with are broken down into the various elements. The main

FIGURE 6.1
Selecting percussive loops in Stylus RMX.

window of Stylus RMX allows you to select loops based on their original tempo. These loops will be played at the tempo of the session, therefore, regardless of what the original tempo is it will play back in time to match the song. There are editing features in Stylus RMX that will allow randomization of various parameters for these loops. The Chaos Designer module allows you to adjust and randomize the pitch and dynamics of the selected loop (Figure 6.2). For each loop there is an associated MIDI file.

Once a loop has been selected, the associated MIDI file can be dragged from the plug-in and dropped into an MIDI track of the digital audio workstation (DAW). This will allow the copying and pasting of the MIDI file for the specific sections for the loop. Stylus RMX does not utilize the MIDI note data as other programs do, by denoting a note on and note off for each percussive element. Rather, it divides the loop up into 16th-note slices. These slices are assigned a specific

FIGURE 6.2
Stylus RMX's Chaos Designer creates randomization of different parameters.

MIDI note, and the MIDI data are a note on for that particular slice. It increments up a chromatic scale to give each slice a unique note number. See Figure 6.3.

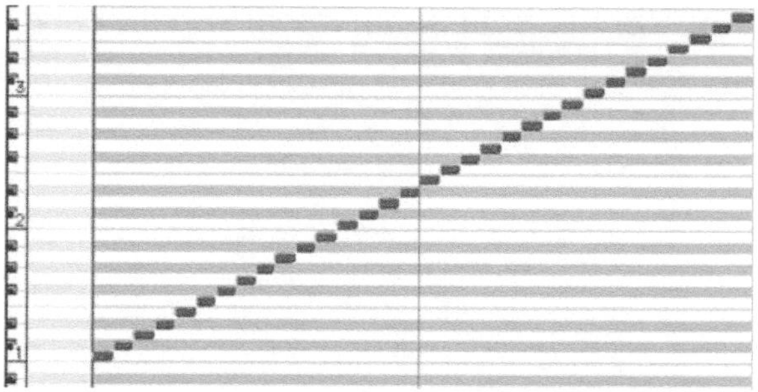

Rhythmic sequences using Reason

Sequenced loops can also come from devices connected through Rewire. Propellerhead's Reason can use several of its devices to create a rhythmic sequence. The Dr.REX Loop Player module can be used to playback loops. These loops can be sequenced either in Reason or the host DAW.

FIGURE 6.3
The MIDI file from Stylus RMX indicating the different slices triggered by each note.

When highlighting a selection in the DAW, the same selection also gets highlighted in Reason. This allows you to select a region in the DAW where you want to place the loop and then paste the MIDI sequence data from the Dr.REX Loop Player directly into the MIDI sequencer of Reason (Figure 6.4). This MIDI data file can then be exported from Reason, imported into the DAW, and then assigned to that device as an MIDI output destination. Since Reason needs to be open anyway, rhythmic sequences are best left in Reason's internal sequencer. Dr.REX uses the same means of MIDI sequencing as Spectrasonics' Stylus RMX, and it has a specific consecutive MIDI note message for each slice of the loop.

Reason can also be used to program your own drum patterns using the Redrum Drum Computer module. This module can load preset groups of sounds, individually selected sounds, or sounds that come from your own sample library. Programming a pattern this way allows the sequencing of the loop to be done while the song is looping over in the appropriate section. With the sound looping, in context with the song, you can hear the contribution of the different elements added from the programmed drum loop. The overall pitch, level, and panning can be adjusted to balance the loop to fit the song. The pattern can then be exported to Reason's sequencer so that it can be copied and pasted to the appropriate sections of the song.

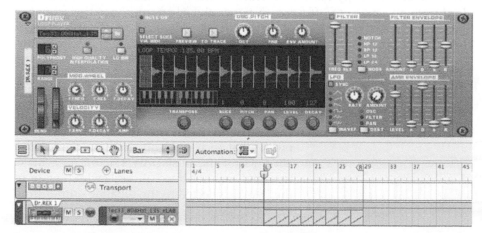

FIGURE 6.4
The sequence from Dr.REX copied into the MIDI sequencer.

FIGURE 6.5
The drum pattern from the Redrum module pasted into the sequencer of Reason.

FIGURE 6.6
Deselecting the "Enable Pattern Selection" button to allow for the playback from the MIDI sequencer.

In order to transfer the sequence from the Redrum Drum Computer sequencer track in Reason to its MIDI sequencer, right-click the Redrum unit and select "Copy Pattern to Track." This MIDI sequence has the standard note on and note off for each percussive element. Once the sequence has been transferred into a sequencer track, deselect the "Enable Pattern Selection" button of the Redrum module so that the sequence data are coming directly from the MIDI track and not the internal pattern sequencer of the drum module (Figure 6.6).

If multiple modules are used with Reason, be sure each one is patched to an individual output so that each one will show up on a different track of the host DAW (Figure 6.7). Make sure that each instance of the Rewire plug-in in the DAW is selected to the separate output from Reason (Figure 6.8).

FIGURE 6.7
The back of the Reason hardware, to create multiple outputs in the host DAW.

ADDING RHYTHMIC MELODIC ELEMENTS

An interesting combination of rhythm and melody is the addition of melodic keyboard parts that have rhythmic elements that are synchronized to the tempo of the song. This can be used for pads that are carrying the chord changes of the song to make them more interesting, rather than just functioning as a standard pad. Most software synthesizers have the ability to synchronize some modulations to the tempo map. These sounds do not necessarily replace a drum loop, but they can add a rhythmic texture to the song. These tempo-synchronized sounds can come from software instrument plug-ins or a software synthesizer that connects via Rewire, as Rewire transmits tempo and song position information. Oftentimes, the individual patches in the software instrument will be categorized inside the software instrument itself, making them easier to find.

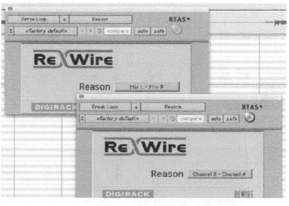

FIGURE 6.8
The different Rewire plug-in instances outputting the different outputs from the Reason hardware.

When trying to find these tempo-sequenced sounds, make sure that the software instrument supports tempo synchronization. Many software instruments that emulate vintage analog synthesizers will not have this type of synchronization, as it was never a feature of the original synthesizer.

These synthesizer sounds can have a variety of parameters synchronized to the tempo. The most common parameter that is synchronized is the low-frequency oscillator (LFO), which can be used to modulate any number of parameters of the synthesizer patch. For instance, it can be attached to the filter, which will create a changing texture, or it can be attached to the amplifier, which will create a more rhythmic sound. In order for the LFO's rhythm to be noticeable, it is usually set to a stepped modulation, such as a square wave, or any number of proprietary LFO patterns.

When sequencing these parts with Reason, the MIDI sequencing can be recorded in the host DAW. Once it has been created, the output of the MIDI track needs to be assigned to the appropriate device in Reason. There are two devices that do this inside of Reason. The first is the Malstrøm—it features a modulator that can be synchronized to the tempo of the song (Figure 6.9). The second device is the Thor polyphonic synthesizer; this instrument features many parameters that can be synchronized to the tempo map (Figure 6.10).

FIGURE 6.9
The Malstrøm synthesizer with the Mod B synchronized to the host tempo map.

FIGURE 6.10
The Thor synthesizer with the delay and LFO 1 synchronized to the host tempo map.

Learning more about music synthesis and the signal flow synthesizers will help to fine-tune the settings of a particular instrument patch to accentuate the synchronization to the tempo map (Figure 6.11).

Once the MIDI data of the sequenced parts have been recorded in the DAW, the tracks need to be quantized at 100 percent to the beginning of each beat. Depending on the performance, it may be necessary to also quantize the durations of the MIDI notes so there is no audible gap when changing chords.

EMULATED INSTRUMENTS

Using emulated software instruments is a way of adding acoustic or electric sounds that would otherwise be inconvenient or expensive to add to a recording.

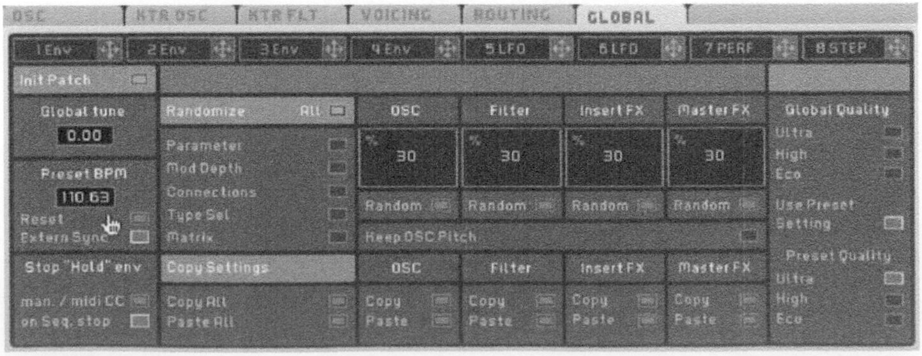

FIGURE 6.11
Massive by Native Instruments, reading the tempo fluctuations from the host DAW.

There are many different software synthesizers available to create any sort of additional instrumentation on a recording. Software emulation of instruments falls into two categories: emulating acoustic instruments and emulating synthesized instruments. In addition to emulations, there are also completely original instruments available with software synthesizers.

Emulated sounds, such as an organ or an acoustic piano, are helpful additions to any studio, as it can become expensive in time, money, and space to have these instruments available and to keep them properly maintained. For faster processors and larger hard drive capacities, there are sample-based instruments that can sample multiple notes in stereo and at different velocities. This can help recreate a natural orchestra or piano sound, which are generally difficult to artificially synthesize with any credible accuracy.

When adding an emulated instrument it is best to understand the instrument as much as possible. When adding an organ emulator it helps to understand how an organ works. These software synthesizers will emulate the controls of the different instruments and it would be helpful to know what the drawbars and presets do to the sound. When working with an emulated synthesizer, it helps to understand the different functions of the instrument being emulated.

As with any software instrument, each one will add additional usage to your computer's processing. The more emulated instruments that are added, the less processing that is left over for mixing plug-ins. If the project has multiple software instruments open prior to mixing, it may be best to bounce those instruments as audio tracks to free up some space on your computer's CPU.

Organ emulation

An organ is one of the most common sounds used in the studio. It can serve to function as a pad, rhythm, or lead instrument. There are a wide range of tones that can be generated with a Hammond B3 organ, which is the most common organ used in recordings. Most small studios do not have the budget or the space to have a Hammond B3 organ with the required Leslie 122 cabinet (Figure 6.12).

Synthesizing an organ itself is not difficult. There are many different simple harmonics that can be added, mimicking the tone wheels inside the organ. The most important feature of an organ emulation is the accuracy of the rotary speaker emulation. This is the sound that has become synonymous with the organ.

The rotary speaker consists of two upper horns and a lower horn that rotate, creating a tremolo and Doppler effect; see, for example, Figure 6.12. The rotary cabinet also has the ability to change speeds from slow to fast. The time that the rotary cabinet takes to make these speed accelerations and decelerations is part of the organ sound. It is the emulation of this rotary speaker

FIGURE 6.12
The upper rotating horns of a Leslie 122 cabinet.

that becomes crucial to the quality of any emulated organ sound. Additionally, vintage rotary speakers have tube amplifiers that can be overdriven to create a distorted sound.

Some organ emulators go beyond the standard Hammond B3, and they will emulate other popular organs such as the Harmonium, Farfisa, or Vox Continental. These different emulations allow the creation of tones similar to those used by The Doors or The Beatles.

Native Instruments' B4 II is a commonly used organ emulator (Figure 6.13). The front panel looks like that of an actual Hammond B3 organ, with drawbars, presets, and dual registers. Additionally, you can go to the expert page, which will allow for more fine-tuning of the instrument. As with all vintage instruments, no two organs will sound exactly alike. There are many different fine-tuning adjustments of the rotary speaker as well as calibration of the organ itself. Moving to the expert page of the B4 II instrument will allow you to adjust the speed at which the rotary cabinet accelerates from slow to fast. See Figure 6.14.

Acoustic piano emulation

In addition to the organ, having an acoustic piano in a studio is a luxury in this day and age. A good acoustic piano is very expensive to purchase and requires regular tunings to make it useful in the studio. An acoustic piano is an instrument that is best emulated with samples rather than synthesis. There are many different software pianos available that are comprised of many different layers of stereo-recorded samples. Samples can be recorded of each note with different velocities to create a realistic piano sound.

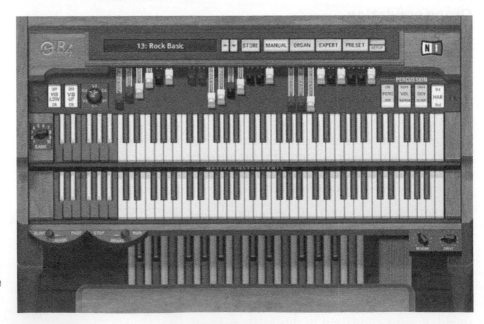

FIGURE 6.13
Native Instruments' B4 II instrument, modeling the controls and layout of a Hammond B3.

FIGURE 6.14
The advanced editing features of the B4 II, allowing modifications of the rotary cabinet parameters.

To the dedicated pianist, this sampled piano may not be suitable for solo work, but in the context of a pop recording, it can be very convincing. Since the standard piano has 88 keys, sampling the full duration of each string with the different velocity layers can take up a tremendous amount of hard drive space. Until the past few years, this quantity of hard drive space was expensive. With today's larger hard drives, it is possible to have many different sample libraries on the same drive. Since many of these sample libraries can be 50 gigabytes or more, getting the largest hard drive you can afford will help you be better off for obtaining and creating additional sound libraries.

When recording a pianist, it is best to have a suitable MIDI controller that feels as close to a piano as possible. A cheap MIDI controller may not have the weighted action that piano players are used to, as well as the familiar 88-key range. It is also helpful to have a sustain pedal for the MIDI controller so the piano player can play with a sustain pedal as if he or she was playing a real piano.

An acoustic piano does not have near the options that other emulated instruments have. There are no knobs or buttons on a piano, only pedals. There can be some adjustments to the sound, such as the amount of reverb, which should be turned off if there is a suitable reverb plug-in available.

An acoustic piano does not necessarily need to be emulated by a dedicated piano plug-in. There are many different sample libraries that will work with different types of stand-alone sample playback plug-ins. The libraries that contain the largest amount of individual samples will generally sound the most realistic. See, for example, Figure 6.15.

FIGURE 6.15
The Akoustik Piano from Native Instruments, featuring multiple sampled pianos and a few parameters to modify.

Electric piano emulation

Emulating an electric piano, such as a Rhodes, Wurlitzer, or Clavinet, are also valuable additions to a studio. The Rhodes piano is a classic sound that was used on many recordings in the 1960s and 1970s, including The Beatles' "Get Back." The Wurlitzer electric piano, which has a grittier sound than the Rhodes, was also frequently used during that time period on records like "Dreamer" by Supertramp. A Clavinet, which has a distinctive plucked sound, was used in many funk records, such as Stevie Wonder's "Superstition."

There are single plug-ins that will emulate each of these different instruments separately, such as Logic's EVP88 and EVD6, as well as stand-alone plug-ins, which will potentially emulate all three, like Native Instruments' Elektrik. Many different sample libraries will have sample banks for these different instruments. See Figures 6.16 and 6.17 for examples.

Analog synthesizer emulation

The emulation of analog synthesizers is common with many software plug-ins. There are dedicated plug-ins that will emulate specific synthesizers. Some may be licensed, such as Arturia's Minimoog V, but others will be general analog subtractive synthesis emulators. Having an understanding of subtractive synthesis, as well as the different software instruments, will help you make the best use of these modules.

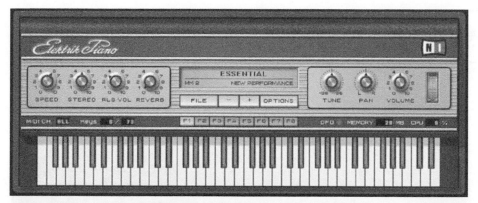

FIGURE 6.16
The Elektrik piano module from Native Instruments, visually resembling a Rhodes piano.

FIGURE 6.17
The NN-XT module from Reason playing back samples from a Wurlitzer electric piano.

Some software synthesizers will go beyond the original functionality of the instrument they are emulating. An example of this would be Native Instruments' FM8. This is originally an emulation of the Yamaha DX7 FM synthesizer. It goes beyond the capabilities of the original synthesizer, and adds more operators, as well as the ability to create your own algorithms (Figure 6.18). For a specialized synthesizer, such as the FM8, it helps to gain an understanding of the principles of FM synthesis, and the way that operator pairs work.

Drum emulation

Emulating drum sounds can be a difficult task. There are many different types of drum sounds used in audio production. There is the emulation of electric drum machine sounds, as well as acoustic drums.

The emulation of electric drums is fairly straightforward. Most drum machines have a limited amount of sounds they can output, especially vintage drum machines. These drum machines are easily re-created by sample playback instruments. There are several libraries of these electric drum machine sounds that can be purchased, as well as some that users post on the Internet for free. Some analog drum machines, such as the Roland 808, are analog drum synthesizers. The pitch and envelope of some of these analog drum machine sounds can be adjusted, so a sample library would need to have many different variations of the individual sounds in order to best re-create the sound of the instrument.

Emulating an acoustic drum kit is far more complicated than that of the electric drum machine. When recording an acoustic drum kit in the studio, every microphone picks up every drum in the room. The individual drums will also resonate when the other drums are struck. This can create a difficulty in just playing back individually recorded drum samples to recreate the sound of a drum kit.

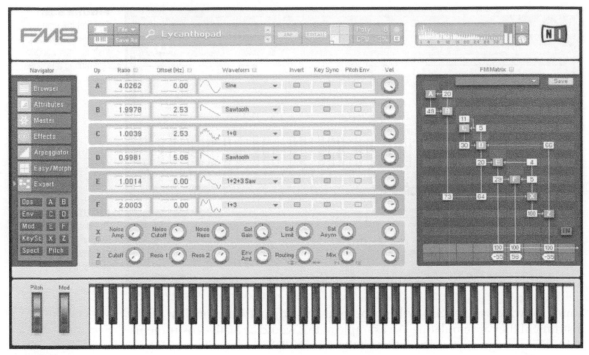

FIGURE 6.18
The FM8 synthesizer that can model the FM synthesis of a Yamaha DX7, but with more capabilities.

In addition to emulating the sounds of a drum kit, the technique that a drummer uses to play the drums needs to be emulated as well. Drums may sound different when they are hit with the left or right stick. They even sound different, depending on where they are struck. A snare drum struck toward the side of the drum head will sound different than if it is struck on the side. The ride cymbal will sound different if it is struck on the main part of the cymbal than if it was struck on the bell. Additionally, the drums will sound different with each hit, so an instrument with just a single sample will have a "machine gun" sound to it when playing back that same sample in succession.

There are currently different software plug-ins that allow for the emulation of both the way the acoustic drums sound, as well as the manner in which the drummer plays the drums. FXpansion's BFD and Toontrack's Superior Drummer and EZdrummer offer a sample-based solution that covers all of these bases. These plug-ins will have multiple samples for each drum hit, as well as different velocity layers, similar to a dedicated sampled acoustic piano. These plug-ins are also expandable with different sample libraries that can be purchased from the manufacturer.

These different drum plug-ins allow for a tremendous amount of control over what is heard on each track. You have the ability to control whether or not the individual microphone will pick up the other instruments, as well as how

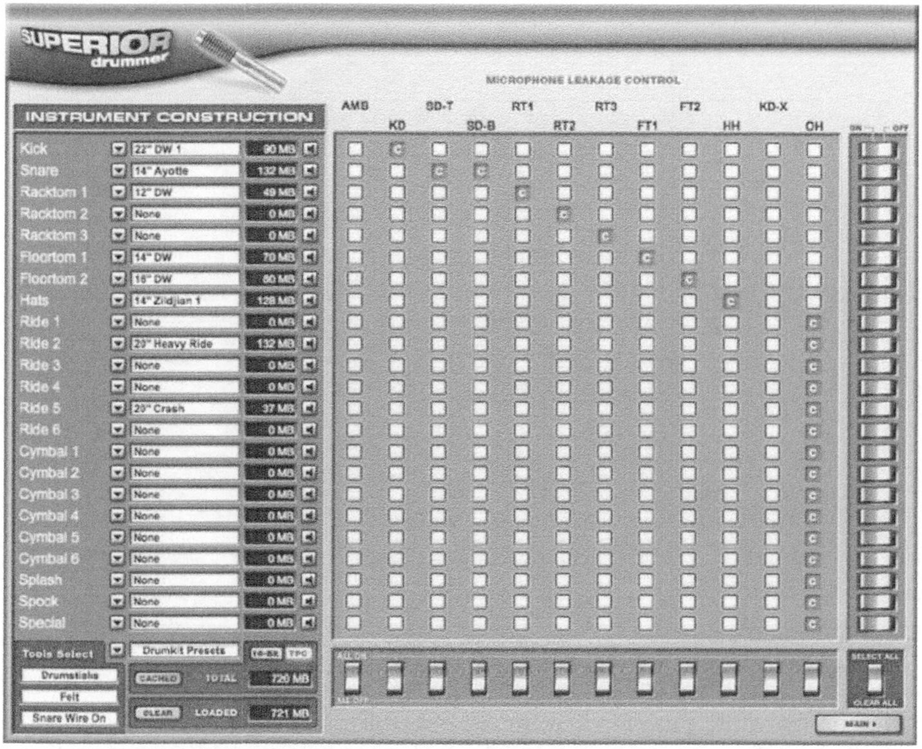

FIGURE 6.19
Toontrack's Superior Drummer adjusting the bleed from the different drum tracks into the different microphones.

the drums will appear in the overheads and ambient room microphones. This helps to create a realistic-sounding drum kit. Since these instruments are sample based, the more you have the drums bleeding into the other microphones, the more memory is required for that plug-in. This will require a computer with a fast processor and expanded memory. See Figure 6.19.

Emulating instruments with a sampler

Utilizing plug-in samplers is a way of emulating any instrument that can be recorded. These can be used to create additional orchestration for a song, adding any exotic percussion sounds, or even playing back sampled synthesizers. The quality of software samplers has increased tremendously over the years, as they have more memory available and hard drive storage.

A sampler can be used to recreate any instrument possible, provided that there are those samples in the library. A sampler can be used to create your own instruments, as well as playback those that were created by others. Software samplers will come with their own sound banks and the ability to import sample banks of different formats. These instruments can vary in the amount of layers and processing they can apply to the sounds. Some software samplers are easier to use, but less flexible, while others are more complicated, but can do a tremendous amount of processing, as well as allow for several layers of samples.

FIGURE 6.20
The Kontakt sampler plays back various sample banks from an orchestra.

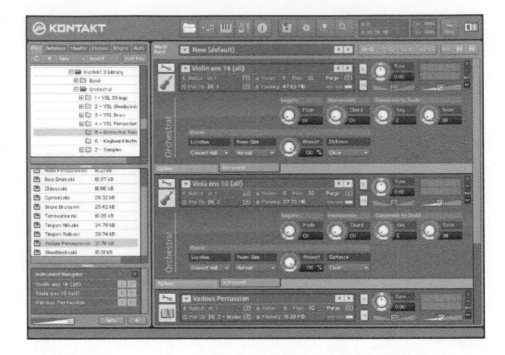

The quality of the sampled instrument depends on the person who created it. Of all the software instruments available, having a large selection of realistic, high-quality samples can be very expensive. If you take, for instance, an orchestra, there are many different ways a string section can play a note. For example, it can be played pizzicato, or it can slide from note to note, so all of these different possible variations need to be captured. The best string sample libraries can cost thousands of dollars. If you are looking to be a film scorer, then this may be a worthwhile investment; however, if you are just looking to add some good string sounds to a pop recording, then the standard samples that come with a sampler may be sufficient (Figure 6.20).

A FINAL WORD ABOUT ADDING MIDI TRACKS TO RECORDINGS

Being able to add a few synthesized parts, whether they are drum loops, emulated instruments, or just a pad to carry the harmonic content, will help expand the tools available in a home or project studio. It is possible to go overboard and overtake the main instruments with the synthesized sounds. Always work with the client to determine how many extra synthesized parts should be added.

CHAPTER 7
Mixing Techniques

All of the work done throughout the recording process leads up to the final mixing stage. Many producers and engineers believe that this is the most important stage of making a record, so it is very important to make sure that you and your client allocate enough time and budget to mix the song(s) properly. Mixing a track can be a time-consuming process. All of the tracks need to be equalized, balanced, effected, and placed in their appropriate position in the sonic spectrum. This is where the engineer will add compression to contain the dynamic range, reverb to give the recording a sense of space, and any other appropriate effects. It can be a daunting task combining many tracks into a single stereo pair. The more tracks that have been recorded, the longer the mixing process will take. Since mixing is an additive process, in that you are combining tracks, many of the steps that are taken are designed to make room for all the tracks.

PREPARING THE SESSION

Throughout the overdub process you may have found yourself adding compression, equalization, and other effects. Sometimes it is best to start with a clean slate. Unless you are married to a particular sound, such as a guitar cabinet simulator, which creates the main sound of a track, you may find that removing all the plug-ins with the exception of any pitch correction, sound replacing, and instrument emulation is a good way to build the mix up from the raw tracks. This way you are not biased in the way that a particular track sounds from hearing the same settings over and over again. Most of the editing, such as timing correction and arranging, hopefully has been done before you start your mix. During the mixdown, you may find that you need to edit some parts, as you will hear that certain parts may not work in the mix as well without some additional editing or pasting.

FIGURE 7.1
Saving the separate mix session.

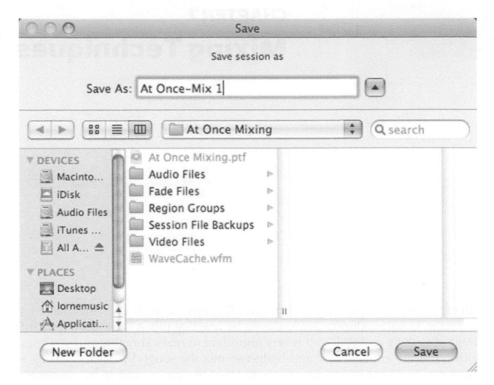

Since mixing oftentimes requires revisions based on listening to the mix in different listening environments, as well as fulfilling the client's wishes, it is helpful to save a separate version of the mix each time you do an update (Figure 7.1). This way you can always go back and reference what you have done previously. For each different mix made of a particular song, it is a good practice to keep track of notes detailing the difference of each version so that you can easily go back and re-create or extract the differences for a newly updated mix. Suppose the client likes the way that the vocals sound in mix one, but is happy with the way the rest of the tracks sound in mix three, you can transfer the settings from mix one into mix three and create an entirely new mix four.

Oftentimes when mixing, you may find yourself bringing levels up more than you bring levels down, so it is a good idea to start with lower levels for all of the tracks. That way you have the flexibility to bring tracks up before clipping the stereo mix bus. If your digital audio workstation (DAW) does not automatically give you a stereo master fader, make sure you start with one so you can monitor the levels throughout the mixing process. You can temporarily group all of the tracks and bring each level down at once by a specified amount of decibels. Once compression is added to the tracks, you will see the overall level start to climb up.

Organizing your tracks

It is important to begin a mix with all of your tracks organized. You will find it helpful to have all the guitars, vocals, and background vocals in the same visible area, as oftentimes, you will be working with tracks of a similar instrument. If there are doubled guitar tracks, make sure they are next to each other so that you can easily copy and paste settings between the different tracks. Another feature of DAWs is the ability to group tracks in such a way that you can adjust the volume level on one track and have it adjust the level of the other tracks in the same group. This is helpful if you want to bring the distorted guitars or drums up or down overall in the mix.

Another helpful feature is the ability to subgroup the tracks into their own stereo output. This has the advantage of increasing or decreasing the volume of those similar tracks in the mix, even though you may have volume automation on the individual tracks. An added benefit to having each of the instruments go through a stereo submix is that you place a compressor across all of those tracks. This is helpful if you want to compress all of the guitars with a single stereo compressor so that the guitars will have a common compression across their dynamic range.

In many DAWs you have the ability to label and name the different output and input buses. This helps to keep track of your different stereo submixes. You can create a stereo submix by selecting the output of each of the drum tracks to be Bus 1 and 2. You can then create a stereo auxiliary input with the input being set to Bus 1 and 2, and the output being set to the stereo mix. If there are multiple files for a particular sound source, such as a bass that may have a microphone and a direct input signal, you can create a mono subgroup. With all these different submixes, you can easily mute entire instrument types with the click of a button. See Figures 7.2 and 7.3.

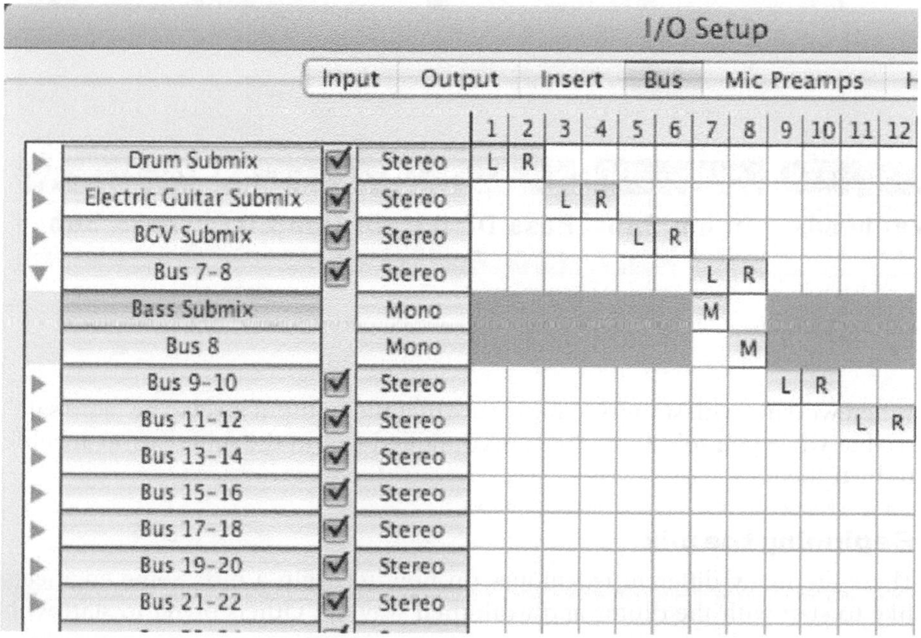

I/O Setup																	
		Input	Output		Insert		Bus		Mic Preamps								
				1	2	3	4	5	6	7	8	9	10	11	12		
▶	Drum Submix	☑	Stereo	L	R												
▶	Electric Guitar Submix	☑	Stereo			L	R										
▶	BGV Submix	☑	Stereo					L	R								
▼	Bus 7–8	☑	Stereo							L	R						
	Bass Submix		Mono							M							
	Bus 8		Mono								M						
▶	Bus 9–10	☑	Stereo									L	R				
▶	Bus 11–12	☑	Stereo											L	R		
▶	Bus 13–14	☑	Stereo														
▶	Bus 15–16	☑	Stereo														
▶	Bus 17–18	☑	Stereo														
▶	Bus 19–20	☑	Stereo														
▶	Bus 21–22	☑	Stereo														

FIGURE 7.2
Naming the separate output of the buses so that the names appear on the outputs.

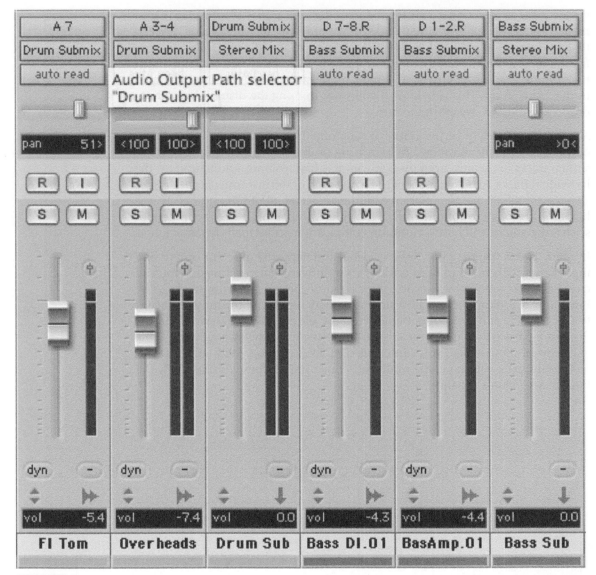

FIGURE 7.3
The name of the buses appear on the submixes.

When working with submixes, make sure that their output is set to be "solo safe" so that when you solo tracks that are contained within the submix, you are able to hear them.

Beginning the mix

There are many different techniques on how to begin a mix. Some engineers like to start with the drums and work their way up. Others prefer to start with the vocals as they are the key instrument to build everything around. There is no right or wrong way to begin a mix; each engineer is different. The most common technique, however, is to begin with the drum tracks. It is a good idea to have a generic level set for all the tracks that are in the song. In this way, you can hear how the tracks you are focusing on work in the context of all the other instruments. In Pro Tools this is easily accomplished by Command-clicking (Mac) or Control-clicking (PC) the solo button.

PANNING AUDIO

Manipulating tracks in the stereo spectrum through the use of panning is a means to build a wide stereo mix. Many engineers when they are starting out make the mistake of recording everything in stereo and thinking this will create a wide mix. This is a mistake, however, because if everything is recorded in stereo, and hard panned left and right, the sound completely occupies the center of the stereo spectrum. If you listen to commercial recordings, you will hear that most of the audio tracks are coming across as mono and panned in a particular location in the stereo spectrum. If there is a particular instrument that is recorded stereo, do not be afraid to either eliminate one of the sides of the recording so that it is a mono track, or pan both tracks into a single mono position.

Certain instruments, such as an acoustic guitar, piano, or some electric guitars, work well as a stereo instrument. This is determined by the style of music. If the focus of a song is on the acoustic guitar, then it most likely should remain a stereo instrument. If acoustic guitar is used as a rhythm instrument, it may be better to have a doubled mono track as opposed to a single stereo track. This gives the engineer the ability to hard pan both mono acoustic tracks, but it also leaves a hole in the center for the other focal instruments.

MIXING DRUM TRACKS

When mixing drum tracks, it helps to have an understanding of what should be present in the mix. When most people think of drums and they hear drums on a recording, the majority of the time the bulk of the drums consists of the kick and snare. The hi-hat carries high-frequency information and subdivisions of the beats. The cymbals are generally used to accent certain beats, as well as to keep time with the ride cymbal. The ride cymbal and the hi-hat often go hand-in-hand in carrying the rhythmic subdivisions of a drumbeat. If the drummer's focus is on the hi-hat, then there will be little ride cymbal played and vice versa. These instruments are separated in the stereo spectrum by their physical placement on the drum kit. The hi-hat is on the drummer's left side and the ride cymbal is on the drummer's right side. Toms can carry the beat during certain sections of the song, but they can also be used for drum fills into the different sections of the song or to accent certain beats.

Before you begin mixing the drums, decide how you want them panned in the stereo field. The two choices are either drummer's perspective or audience perspective. With drummer's perspective, the drums are panned as if the listener were the drummer. If the drums are panned with audience perspective, the drums come across the stereo spectrum, panned to where the listener would see the drummer play, as if he or she were standing in front of the drum kit. There is certainly no right or wrong way, and as you listen to commercial recordings, you will hear both panning styles being used. The important thing is to be consistent. If the drums are panned to the drummer's perspective the tom fills will go from left to right. The hi-hat will be on the left side and the ride cymbal on the right. Make sure the overheads are also panned accordingly. The drums are panned oppositely if the drums are panned to the audience's perspective.

When putting plug-ins across the drum tracks, make sure that you place them in the optimal order. If there is any gating to be done, make sure that this is the first plug-in across the track, with the exception of any high-pass filtering. Otherwise, a compressed dynamic range will make the drum track more difficult to gate after the compressor. Equalization is generally next in the signal chain. One of the advantages of using plug-ins over outboard equipment is that you have the ability to save commonly used presets. These presets can give you a great starting point and can be useful for making some subtle adjustments to tailor the sound to the recorded material.

Kick drum

When equalizing a kick drum, the focus is usually on the attack as well as the low end. You can focus the equalizer in on the attack by creating extreme boost on the high frequencies and adjusting the frequency in such a way that you hear more of the attack from the kick drum (Figure 7.4). This equalizer can then be

FIGURE 7.4
An equalizer placed on the kick drum, accentuating the attack and subfrequencies.

adjusted by lowering the boost to where it sounds best for that drum. The key low frequencies of the kick drum are usually around the 100-Hz vicinity. You can use the same process to find where to boost the low frequencies in order to get the most thump out of the kick drum. The low to mid frequencies between the attack and low thump can be reduced, as there is not much viable sound in this region to define the kick drum. Make sure that with any equalizing you do while mixing, you A/B the equalizer in and out to make sure you are achieving the desired results.

Snare drum

When working with a snare drum in a mix, you need to decide whether to gate the snare drum. Applying a gate across the snare drum can eliminate some of the background noise that may be brought up when adding compression. If the drummer is playing ghost notes on the snare drum, then gating may be impossible, as the gate will remove the ghost notes. If the drummer does not play ghost notes throughout the entire song, you can use a gate in the sections where there are no ghost notes.

There are two methods for incorporating a gate on a snare drum track with ghost notes (Figure 7.5). The first is by duplicating the track and deleting the respective regions on each track so that the original track uses the gate, while the duplicated track contains only the audio regions with the ghost notes and has the gate removed. The other, easier method to use a gate would be to automate the gating plug-in so that the gate is only active during the sections without ghost notes. When applying a gate to the snare drum, make sure that you have the fastest attack setting available or some of the transient attack will be lost. The hold and release times should be set in such a way that you hear the body of the

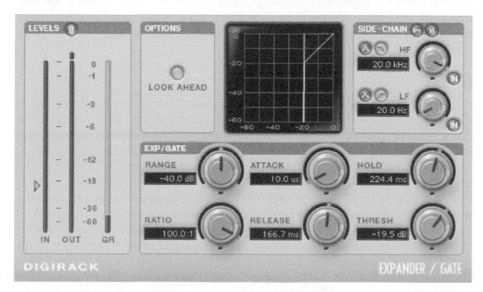

FIGURE 7.5
A gate placed across the snare drum to eliminate background bleed.

snare drum, but it decays enough before hearing any substantial sound from the drums in the background, such as the kick drum.

Equalizing the snare drum can be achieved by bringing out some of the body of the drum as well as some of the attack. If a dynamic microphone was used on the snare drum, you may need to boost the high frequencies to get the sound of the snares to come through. A condenser microphone will capture the high frequencies better than a dynamic microphone will, so you may not need to use as much high-frequency equalization as you would with a dynamic microphone. If you want more body from the snare drum, a boost, around the 250-Hz vicinity, can make a fuller-sounding snare drum.

Snare drums are one of the more frequently compressed instruments during a mix. You can create a flatter-sounding snare drum with less attack but longer sustain by using a compressor with a fast attack and release time. You can accentuate the attack of a snare drum by having a longer attack time and longer release time. This will allow the attack to come through uncompressed, while the decay of the snare drum is compressed. As with equalization, you should A/B the compression to make sure that the compression works for the sound in the track.

If there is a snare-bottom microphone, then it can be gated similarly to the snare-top microphone. The only difference is that there may be more bleed from the other instruments in the snare-bottom microphone, which can make gating the snare-bottom microphone more difficult. One way to accurately gate the snare-bottom microphone is to key it from the snare-top microphone. This can be done by using an auxiliary send, prefader to the key input of the snare-bottom gate. This way, the gate will be acting on the input received from the snare-top microphone as opposed to audio input from the snare-bottom microphone (Figure 7.6). If there is considerable bleed from the kick drum into the snare-bottom microphone, consider placing a high-pass filter on the snare-bottom track in order to filter out the kick drum. There is not much low-frequency information coming from the bottom snare drum, so this is an easy way to clean up the track.

Hi-hat

Mixing the hi-hat microphone as well as any other additional cymbal tracks is different from the kick and snare tracks. The sound you hear in a mix of the hi-hat is generally not entirely from that microphone. The overhead microphones pick up the hi-hat as well. When soloing the hi-hat microphone, any equalization that you do will not represent the way that the cymbal will sound in the mix. The individual track will need to be balanced with the overhead tracks as well to get the optimal hi-hat sound. Since the hi-hat consists of mostly high frequencies, a high-pass filter can be placed across the track to eliminate any bleed from the kick or other drums. Careful equalization of the hi-hat microphone can be done to bring out the crispness of the track and make it stand out in the mix. If there is substantial hi-hat in the overheads, then bringing up the hi-hat track can make it sound overpoweringly loud in the mix. Juggling the sound of

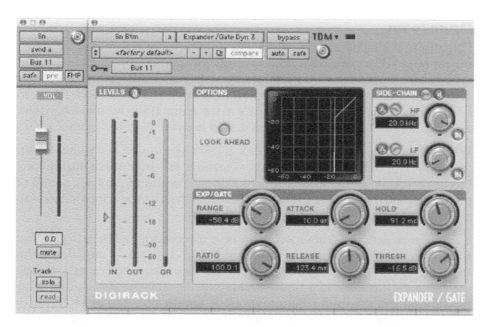

FIGURE 7.6
Snare-bottom
microphone keyed off of
the input from the snare-
top track.

a single cymbal in multiple tracks can be a delicate balancing act, and reflective
on the microphone placement during basic tracking.

Toms

Oftentimes in a song, there may not be very many tom hits. This is where working in a DAW has a distinct advantage over its analog counterparts. You can loop
the sections with the tom hits in order to accurately equalize them. Toms can
have a very full sound, with a solid attack and substantial low end. These can
be accentuated with the equalizer. Boosting the high attack and low body of
the toms, as well as eliminating some of the midrange of the sound, can make
them shine in a mix. Toms can also be compressed, either separately or together
through a stereo submix.

Overheads

The overhead microphones are primarily there to capture the cymbals.
Depending on their placement in proximity to the cymbals, the other drums in
the room may be captured louder in the overhead microphones than the actual
cymbals. A high-pass filter placed across the overhead tracks can help eliminate
much of the room sound and allow you to hear the overhead tracks louder in
the mix without bringing the other drums up. Just because cymbals consist primarily of high-frequency content does not mean that all the focus should be
on these frequencies. If the high frequencies are boosted too much, in combination with a high-pass filter, then the cymbals can come across as sounding
too thin.

Drum room

Oftentimes during basic tracking, there is a drum room microphone. A stereo microphone setup can also be used to create an additional width, but a mono track is used most often. This microphone can be either disregarded completely or incorporated as part of the ambient drum sound. The drum room track can be used to give the drums a sense of space in the room. If you are looking to have a cleaner drum sound in the mix, then do not hesitate to mute this track.

When using a drum room track in a mix, it can be processed in a couple of different ways for different desired effects. Generally, the drum room microphone has heavy compression placed across it. This accentuates the ambient sound of the drum room microphone. You can then equalize the track appropriately to create an added ambience to the sound.

A useful trick with the drum room track is to compress it and then to gate that track with the triggered input of the snare drum. This utilizes the drum room track as a means of creating a fuller, more ambient snare drum sound. Gating the room track is done in a similar fashion as the snare-bottom microphone. Just add the gate across the drum room track and then set it to be keyed off an auxiliary send from the snare-top microphone, which should be sent prefader. The hold and release times can be adjusted to accommodate the sound and tempo of the song.

Drum reverb

Most mixes will have a separate drum reverb set up for certain drum tracks. Not every drum track has reverb applied across it. There can be reverb that is only applied to the snare drum. Oftentimes, toms are added to the reverb well. Different ambiences, as well as plates, are used for the sound of the drum reverb.

The most important setting when creating a reverb for the drum tracks is a decay time. This is adjusted to accommodate the drummer's performance as well as the speed of the track. You will want to make sure that the decay is mostly completed before the next drum hit, otherwise the sound of the drum tracks will be muddied with the added reverb. Reverb is generally not added to the kick drum track unless you are going for a specific sound for that track. The kick drum track will lose some of its definition with added reverb. Adding reverb to overhead cymbals essentially adds a reverb to everything, as the cymbal tracks record every drum in the room. Reverb can be added to the hi-hat track to make it a little more ambient in the mix.

Bass

When working with the drums in the mix, the next tracks that should be added are those from the bass, as the bass instrument functions rhythmically with the drums. There can be multiple tracks on the bass instrument. There can be a direct input as well as a microphone on the bass cabinet. When mixing, you need to decide whether to use one or both of these tracks. The direct-input sound will have more attack from the bass. The microphone on the bass cabinet will contain more grit to the sound and potentially more low-frequency information.

Since the kick drum and bass carry most of the very low–frequency information, some equalization can be done to make sure that these instruments are not fighting for space in the frequency spectrum. Since the kick drum cuts through the mix at 100 Hz, adding a slight dip to the equalization in the bass track will make more room for the kick to come through the low-frequency range (Figure 7.7). If you are looking to hear more of the attack of the notes, as well as the picking or fingering of the bass notes, a boost in high frequencies where the attack is present can create more definition and differentiation between the notes. Depending on the bass sound that you are trying to achieve for your mix, you may find that you want to reduce the attack frequencies. This all varies depending on the track as well as the musician and instrument itself.

The bass can be compressed in a couple of different ways. Since the natural performance of the bass can be very dynamic, a compressor is almost always applied to the bass. You can use a standard single-channel compressor to compress the bass, generally with a 4:1 ratio. The attack time can be adjusted to control how much of the initial attack of the bass you want to come through the mix. The longer the attack time, the more the attack from the bass is going to come through. Instead of using a standard compressor across the bass track, a multiband compressor can be used to compress only the low end. This allows for some variation on the attack, while helping to create a consistent low end.

FIGURE 7.7
Snare-bottom equalizer set to accentuate the attack and dipped to make room for the kick drum.

GUITARS

If your mixing style is to gradually build your mix up from the bottom begin-ning with the drums and bass, the next step would be to add the rhythm guitars. There could be several guitar tracks in a song, so it is best to start with the most crucial elements of the rhythm performance. Some songs may be acoustic guitar driven and some songs may be electric guitar driven. If the main element of the song is acoustic guitar, then begin by adding these tracks. A song may be piano or keyboard driven, in which case these should be the next instruments added to the mix, before the guitars.

Distorted guitars

Working with distorted electric guitars is different from clean electric guitars. Most distorted guitars and a commercial recording are not as dense sounding since they are captured in the studio. The distorted guitars generally carry more low-frequency information than a clean guitar track. Distorted guitars will also not carry the attack that a clean guitar will. The distorted guitars maintain a more consistent level throughout the performance, however, compression can be used to warm up the sound, as well as to level the dynamic range. Equalizing these distorted guitars can consist of removing some of the low- to mid-range frequencies to make room for other instruments, as well as adding more edge if necessary. A high-pass filter can be useful on a distorted guitar in order to elim-inate any unnecessary low-frequency information. These guitars can be com-pressed slightly to maintain a dynamic consistency.

Clean guitars

A clean guitar track is much more dynamic than a distorted guitar track. Equalization can be done to bring out the attack of the clean guitar, as well as fatten or thin the guitar as required by the mix. Any compression added to the clean guitars can be used to create a consistent dynamic range. Lengthening the attack time of a compressor will allow more of the attack to come through, and shortening it will remove some of the attack, depending on your needs.

You may also choose to add additional effects to the clean guitar parts. Depending on how they function in the song, adding a clean delay to the guitar track can give it some depth. A slight chorusing to the sound can also create a thicker clean guitar sound.

Acoustic guitars

Working with acoustic guitars in a mix with several tracks requires some work with an equalizer to get them into their appropriate space. The actual acous-tic guitar instrument is usually designed to be a very full-sounding instrument. Acoustic guitars are generally played by a musician as a solo instrument, so they have a very wide–frequency range. Equalizing an acoustic guitar track to fit in a mix generally involves thinning the acoustic guitar sound so that it fits better in the song. Oftentimes, acoustic guitars are recorded in stereo to capture the wide sound of this instrument. If you are working with a stereo acoustic guitar track, and the part is doubled, you can feel free to eliminate one side of the acoustic guitar, preferably the microphone part that has the least amount of sound from the strings.

KEYBOARD PARTS

Working with different keyboard parts in a mix requires just as much effort in equalization and compression as the other tracks. Synthesizer sounds are generally very dense, as that is the sound that helps sell the synthesizer. The tracks need to be equalized to fit within the mix, which usually means carving out some of the dense overlapping frequencies in the low and mid ranges. There can also be many dynamic elements for different keyboard synthesizer sounds. A single-band or multiband compressor can work well to contain the dynamics of these synthesizer tracks. Most MIDI sounds are designed to be stereo, and so there can be a lot of stereo movement in these tracks. As with any stereo track, you can make the elements more functional in a mix by summing them into a mono track. These tracks can be panned around without occupying too much space in the stereo spectrum.

If the MIDI tracks you are using are coming from software instruments, be sure to turn off any effects processing that is not necessary for the actual sound. This can further clutter the mix with inferior-sounding effects, such as a bad reverb or compressor. The separate plug-ins in your DAW will work much better for reverb, equalization, and compression.

LEAD VOCALS

The lead vocals are the central focal point in almost every pop recording. Even though the lead vocals may consist of only a single track, there should be ample time spent working with them to make sure that they sit just right in the mix. Any number of effects and plug-ins can be applied onto the lead vocals. Of course there is a standard equalization and compression, but there can also be echo, pitch shifting, chorusing, and any other effect under the sun. All of these other special effects can be used subtly to make the lead vocals stand out in the mix. If there are too many effects placed across the vocal track, the sound can be overly processed and unnatural. There can be times when you may want this processed sound, but in most situations it can be distracting.

Equalizing vocals

There are many things the engineer can do using equalization to enhance the vocals. The first step is to determine whether there are any unnecessary rumbles in the microphone. Depending on how and where the vocals were recorded, there can be excess noise from the control room, air-conditioning, etc. A high-pass filter can eliminate a lot of this unnecessary low-frequency rumble (Figure 7.8). Depending on the lead vocalist, you can usually safely filter out anything below 75 Hz.

FIGURE 7.8
A high-pass filter to
eliminate any excessive
pops.

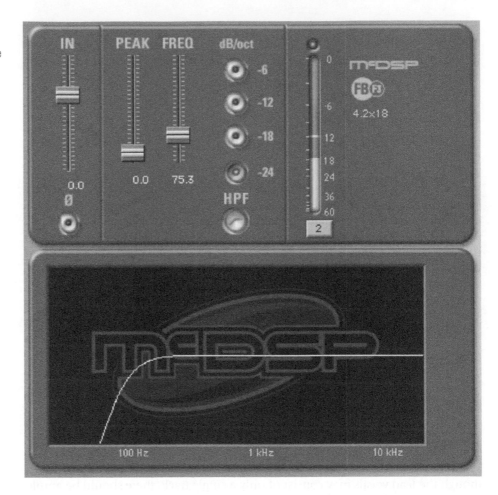

FIGURE 7.8
A high-pass filter to
eliminate any excessive
pops.

Equalizing the vocal track can consist of adding more air to the vocals. A high-frequency shelf that is boosted can put more air and breath back into the vocals. The rest of the equalization is determined by the sound of the vocal track. Sometimes there will be a range of the vocalist's sound that can come across as a piercing sound in the mix. You can utilize a parametric equalizer to reduce these frequencies. You can warm up a vocal track by boosting some of the low range where the vocalist's fundamental for frequencies are.

Compressing vocals

The lead vocals are also the most compressed track in the recording. Compression of the vocals needs to be done carefully so there is no audible pumping and breathing of the compressor. Lead vocals can have a tremendous amount of dynamic range. The vocalist can have verse sections that are much quieter than the choruses. There are two main methods for dealing with this dynamic range discrepancy. The first is to create a separate track that will function as the louder

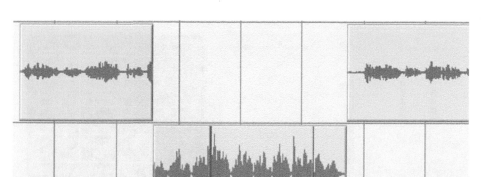

FIGURE 7.9
The vocal track separated into two tracks, one for the softer sections and one for the louder sections.

lead vocals track. This newly added track will consist of the choruses where the lead singer is much louder, so the compressor can be set for those vocals only. The original track can have its compressor set to control the dynamic range of the softer sections. See Figure 7.9.

The other method is to use two compressors in series with each other. The first compressor is used to control the dynamic range of the louder sections. A second compressor is placed after the first, and will then be able to control the dynamic range of both sections, because the louder section is much closer to the softer section in perceived volume after the first compressor.

Removing sibilance in the vocal tracks

After running the vocals through an equalizer and compressor, the added signal processing, along with the vocalist's microphone and technique, may have accentuated the sibilance of the vocal track. Sibilance is the sound of "s" and "t" being overly present in the vocal track. These are high frequencies that range from 3 to 10 kHz. Usually, they are present in a single high-frequency range depending on the vocalist.

Equalization can be used to eliminate sibilance; however, the equalization will also remove the same high frequencies when the sibilance is not present. This is where a specialized effects processor known as a de-esser is valuable since it is designed for situations such as this. A de-esser is specifically created to eliminate sibilance. It is designed to function only when the sibilance crosses a certain threshold. Once the sibilance has crossed the specified threshold, the de-esser will either compress the track or the high frequencies to eliminate the sibilance (Figure 7.10). The method in which a de-esser functions varies by the manufacturer.

Dialing in a de-esser requires precise adjustment so that the singer does not come across as if he or she has a lisp. Most de-essers will have a setting where you can listen to the detection. This is helpful when trying to find the specific

FIGURE 7.10
A de-esser compressing when detecting sibilance.

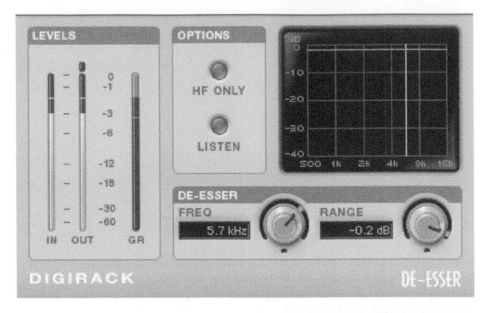

frequency of the singer's sibilance. Once the frequency of the sibilance has been found and selected, then you can adjust the amount of reduction where the sibilance is present.

Effects on lead vocals

Most vocal tracks require some reverb to make the vocals sound more natural. People are used to hearing vocals, even spoken vocals, in a room with reflections. Dry vocals are what is captured in a microphone and sound somewhat unnatural. Adding a reverb to the vocals will put them in a specific space. It will smooth out the vocals and help them blend into the track better, especially if there is added reverb on some of the tracks in the mix. There are other effects that can be added to the vocals to make them stand out more in the mix and give them a better sense of depth.

Reverb on vocals

When selecting a reverb for the vocals, determine what type of reverb best suits the vocalist and style of song. The amount of reverb on vocals varies over time. In the 1980s, most of the instruments had a lot of reverb on them. However, in the past decade, vocals have become much drier. Having a drier vocal sound does not necessarily mean that there is any reverb on it; it can just be a much shorter reverb and have a lower level in the mix. Currently, there is a variety as to how much reverb is on the vocals.

If your DAW has the capability of opening up multiple reverb instances on top of the plug-ins that you are already using, it is best to have a vocal on its own reverb. With the lead vocal track sent to its own reverb, you can tailor the type and duration of the reverb to suit the vocals without having to worry about how the reverb parameters are affecting the drum ambience. A separate reverb will also put the vocals in their own space as opposed to the same space that the drums are in. A short plate style of reverb can help smooth out the vocal sound without giving it the sound of a particular room. A small club setting will put the vocals in a specific room.

When selecting a reverb plug-in to place across vocals in your DAW, be sure that you are using a send for the reverb as opposed to using the reverb plug-in as an insert. This will give you the flexibility of adding the background vocals, or doubled lead to the same reverb as the lead vocals if you choose. A send can be created by using an auxiliary send from the vocal track and putting that send through an auxiliary track with the reverb plug-in set up as an insert. The reverb plug-in should be set to output 100 percent of the wet mix (Figure 7.11).

Most reverb plug-ins will have great presets to help you get a starting point with your reverb sound. However, some of the parameters will require an adjustment for that preset to work for the mix. The width and early reflections can be

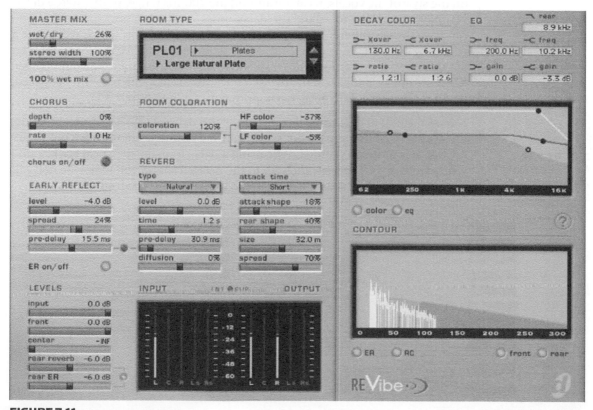

FIGURE 7.11
A reverb plug-in with a plate setting for the vocals.

adjusted, but the most important parameter is the decay time. The decay time should be short enough so that you can hear the decay in between vocal lines. It should be long enough for the vocals to sound smoother. The slower the song, the longer the reverb decay time usually is.

Vocal delay/echo

A delay or echo can be added to the vocal track to create a sense of depth. This has the advantage of overutilizing reverb for depth, as it will not make the overall tracks sound as washy. Delay is often added as an extra send in addition to the reverb that is already placed on the track. Many different engineers have their own tricks as to how much delay they add to the lead vocals.

The difference between a delay and an echo is that an echo will have frequency adjustments to the repeated sound. The high frequencies can be attenuated with a low-pass filter at a specified frequency. Most delays have the ability to adjust this low-pass filter.

There are two main ways to use a delay on vocals. The first is to create a short delay to add some wetness to the vocals. This will have a short delay time. Depending on the style of the vocals, it can be around the 200-ms range. With a stereo delay, you have the ability to adjust the delay times on the left and right sides. You can have slightly different delays on either side of the vocals, which will create a widening effect to the vocal sound. These delays should have the low-pass filter engaged anywhere around the 4-kHz vicinity so that the transients and consonants are not as audible on the delayed track. Without a low-pass filter engaged, it will make your subtle delay track more noticeable (Figure 7.12).

FIGURE 7.12
A stereo delay on the vocals to create a subtle stereo effect.

A good method to determine the amount of this delayed signal to maintain in the mix is to set it so that you do not specifically hear the delay in the track but notice it when it is muted. Even if you are using a delay to make the vocal track sound wetter, adding the same vocal reverb to the delayed sound will make the delay fit in better with the song, as it will be in the same space as the lead vocals.

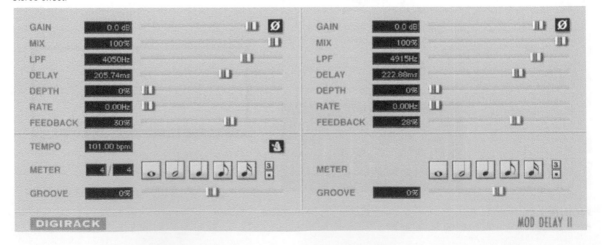

The other method of delay for use with the vocals, which can be used in conjunction with the previously mentioned delay, is to set up an echo of a particular vocal word or phrase that you may want to stand out in the mix. These echoes should repeat in time with the song. The repeats can be set for quarter notes, half notes, whole notes, etc. Most delay plug-ins give you the ability to select one of these musical durations for the delay time. The plug-in then adjusts its delay time according to the tempo from the host DAW. If your plug-in does not support the selection of rhythmic subdivisions as a delay time, it can be easily calculated. Just divide 60 by the beats per minute, multiply that number by 1000, and that will be the delay time for a quarter note in milliseconds. After you have the quarter-note delay time, you can double it in order to get the time for a half-note delay, or halve that number for an 8th-note delay. See Figure 7.13 for an example.

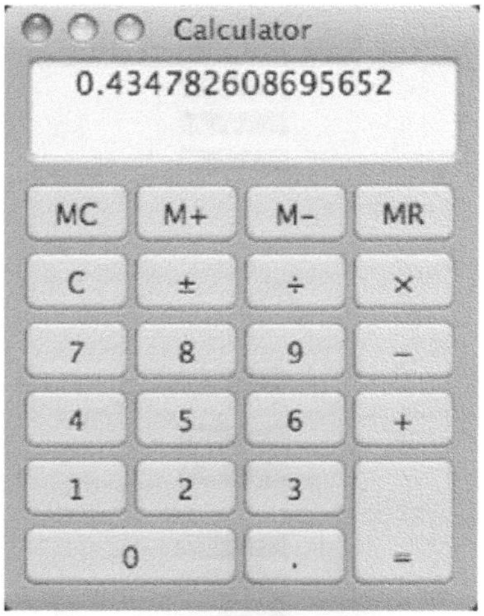

FIGURE 7.13
Quarter-note delay calculated by dividing 60 by the beats per minute, indicating a 435-ms delay for a tempo of 138 beats per minute.

Once the echo has been created and placed in a new auxiliary input, the send feeding that echo can then be automated. Begin by muting this send to the echo. You can then automate the mute of the send to be turned on and off at the locations during the vocal track where you want to hear the echo (Figure 7.14).

Widening vocals with a pitch shifter

Adding a pitch shifter to vocals has been a technique used for years to create a thicker sound. This pitch shifter does not necessarily change the pitch dramatically—only by a few cents. Creating a send from the vocal track to a pitch shifter

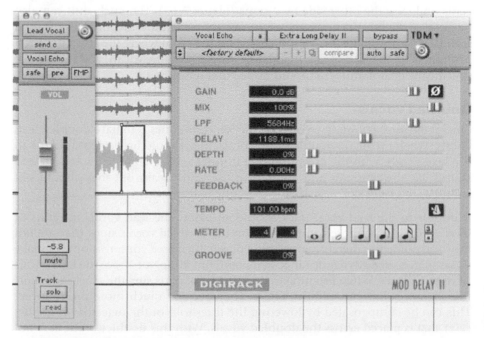

FIGURE 7.14
An echo with its send automated for a single word.

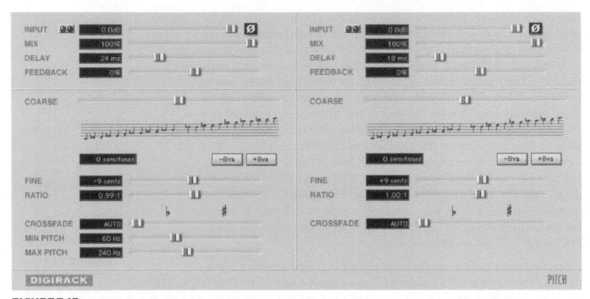

FIGURE 7.15
A pitch shifter added to the lead vocals with ± 9 cents on either side along with a slight delay.

and then adjusting the pitch slightly on either side can create a wide sound. The pitch shifter can make the vocals stand out more in a mix but should be used subtly so that they do not sound artificial. Vocals can be pitch shifted anywhere from 1 to 20 cents on either side. For instance, you can lower one side by 9 cents and raise the other side by 9 cents (Figure 7.15). The pitch shift can also be delayed slightly, in the range of 20 ms, to create more depth to the shifted sound. Do not add the pitch shifting beyond the point where the effect becomes noticeable on the lead vocals. Again, like the delay on the vocals, you should not specifically hear it, but you should notice that it is missing.

Doubled vocals

One of the most common record production techniques is to have doubled lead vocals. When mixing in doubled lead vocals, they should be treated similarly to the lead vocals, with the exception of a thinner sound. Doubled vocals should sound like they support and are part of the lead vocals, but they should not sound as if the vocalist was singing the same part twice.

A double to the lead vocal track should sit below the lead vocals in terms of level. There can be more compression on the doubled vocals since they are not the focal point, but merely a supporting player. Increased compression will make them sit in a single dynamic range better and not jump out above the lead vocals in certain places. When listening to doubled vocal parts sing the exact same line, the sibilances have an additive effect. They become much more pronounced. This can be compensated by lowering the threshold of the detector on the de-esser that is placed across the doubled vocals. With this doubled track, it is not

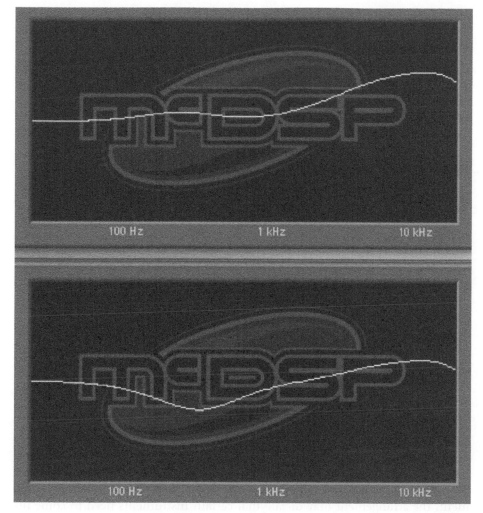

FIGURE 7.16
The upper equalizer is on the lead vocals, and the lower equalizer is on the doubled part with low to mid frequencies dipped.

going to be noticeable if the doubled track sounds a little bit lispy; however, it will be noticeable when there are very strong "s" and "t" sounds.

Equalization on the doubled vocal track can be thinner in the low to mid range (Figure 7.16). This will prevent the lead vocals from becoming noticeably overly dense, because many times the entire lead vocal track might not necessarily be doubled, rather only certain sections of the song.

Harmony vocals

Harmony vocals are a vocal track in which the vocalist sings the harmony part in time with the vocal melody. This functions similarly to the doubled vocal track, and as such, the harmony vocals can be processed in much the same way. The harmony vocals are there as a supporting role, rather than the point of focus.

These harmonized tracks can be compressed the same as the lead vocals, but thinned out with the equalizer so that the vocals do not suddenly come across as being too dense.

There should be at least similar auxiliary sends from the harmonized vocal part as there are on the lead vocal track. You should have at least the same amount of reverb as on the lead vocal track so that it blends. It does not necessarily have to have the same delay or pitch shifting as the lead vocals may have, but at least must be placed in the same acoustic space.

Background vocals

There can also be background vocal parts that do not specifically harmonize with the lead vocals but create a part of their own. These background vocal tracks can be treated as their own instrument. They work with the lead vocals but do not specifically contribute to their sound. Many times these background vocals consist of "oohs," "ahs," or a separate melody from the lead vocals. These backup vocals can be placed in a separate space from the lead vocals by using a separate reverb. This reverb can have a longer decay time and be in a slightly bigger room in order to make these background vocals sound as though they are physically behind the lead vocalist. There can be multiple tracks of these background vocals in order to make them sound more ethereal. These tracks can be compressed and equalized so that they have a consistent dynamic range yet are thinner than the lead vocals sound. If there are many tracks of these background vocals, it may be a good idea to bus them all through a single submix so that they can be compressed together and use a single reverb send from the submix.

AUTOMATION

Automation in mixing is used to create consistent levels with different instruments, as well as to accentuate parts during certain sections of a song. Even though compressors may be used to control the dynamic range of an instrument, the arrangement may dictate that certain instruments need to come up or down during particular sections of a song. A rhythm electric guitar that may seem loud enough during a softer verse may not be loud enough once more instruments come in during the chorus. Automating a vocal track will help bring up softer phrases and the ends of notes as the singer's volume can get buried in the mix. Automation can extend beyond levels and panning to where you can automate parameters of the different plug-ins.

Before beginning automation, make sure that the levels for all the tracks are in a good general position. It is best to set the levels of the faders for where they will be at for most of the song. In Pro Tools and other DAWs, you can choose to the view the track's volume in the Edit window.

Automating sections

Entire sections of a track may be automated up or down without having to go through the entire process of writing the automation with the fader (Figure 7.17).

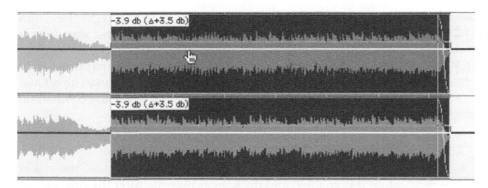

In Pro Tools, it is easy to just separate regions where you want to automate a change in the Edit window so that you can bring the different regions up or down individually. Double-click in the region where you want to adjust the volume, and select the Grabber tool. While looking at the volume level of the track you can move the volume line up with the mouse, and Pro Tools will display the starting level as well as the difference of your volume adjustment. This technique comes in handy when you are trying to bring specific guitars up or down during the choruses.

Fader riding automation

The fader riding method of automation, borrowed from analog consoles, is where you ride the fader, and the fader will then be automated to move according to your fader adjustments that can then be recorded.

There are different types of automation in DAWs as well as their analog counterparts. Before you are able to write any automation with the fader, the automation needs to be set to be recorded. In Pro Tools the default for the automation is to be read. This means if you draw any automation in the volume track the fader will respond accordingly. The other modes of automation vary in what they do to the fader once you are playing back the audio.

Touch automation

The most common method of automation in mixing is touch automation. This will only record automation data when you touch the fader with your mouse or control surface. It will stop recording once that figure is released, and then it will continue playing back any automation that is on the track. If there are no other automation data on the track, it will snap back to its original starting position. Once touch automation is selected, the automation status will be lit red, which indicates that it will be recorded. With the automation adjustments that you make using touch automation, you can easily make adjustments by rerecording the automation data. This is similar to overdubbing a part, only with automation data as opposed to audio data. If there is only one word that needs adjustment, only touch the fader during that word and make the appropriate fader moves. Then release the fader, which will prevent the DAW from writing over previously recorded automation data. See Figures 7.18 through 7.20 for an example.

FIGURE 7.18
The track's automation is set to "touch," which will record when the fader is moved.

Once all the automation data have been recorded to the track, turn off the touch automation and switch it to read. This way there is no accidental automation recorded if you happen to grab the fader while the song is playing back. If you need to listen to the track without the automation, then switch the automation to be off and the fader will then move freely regardless of where the automation is set.

Write automation

Write automation will constantly be recording on the tracks when it is enabled and the song is playing. After the first pass of recording in Pro Tools, this mode automatically reverts to latch mode. The write mode of automation is not as useful as the touch automation, as you will be erasing previously written automation data regardless of whether or not you move the fader. This method of automation is useful if you want to apply automation data across many tracks at once and move the faders on a control surface as if they were a mixing console. This can give you a good initial first pass of automation; otherwise, it is best to stick with the touch automation.

Latch automation

Latch automation functions similarly to touch automation. The only difference is that once you touch the fader and make any automation moves, that fader will always be writing as long as the song is being played back. This is handy in situations where the last level of the fader is where you want it to stay as long as you allow the song to play back.

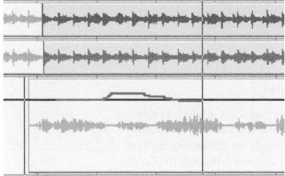

FIGURE 7.19
The automation is being recorded when the fader is moved.

Trim automation

Trimming automation is a means of adjusting the automation data as a whole. While in trim mode, you can raise a fader so it increases the volume of the automation data by 1 dB. This maintains any previously written automation data, but increases the overall level by the amount that you raise or lower the fader. This can be very useful if you have already made detailed automation moves, but a section as a whole needs to come up by a specified amount of decibels.

FIGURE 7.20
The automation after it has been recorded.

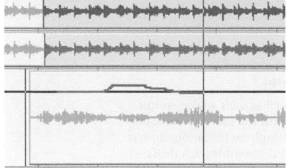

Trimming can be done manually if there are certain sections with automation data that you want to bring up and down. This can be done in the same way that the automation of sections was done as previously mentioned: Double-click with the Selector tool inside a separated region that contains the data that you wish to trim, then switch to the Grabber tool and raise it up or down the amount that you wish to trim the automation data (Figure 7.21). You may need to add a break point at either end of the region, as Pro Tools will only trim automation break points and not the actual volume line itself. This can be accomplished by using the Grabber tool and clicking on the volume level line where you wish to add a break point.

FIGURE 7.21
A section of the vocal, with automation recorded, being trimmed down 3.5 dB.

COMPLETING THE MIX

Even though you may go through and work on the mix track by track, there are constant adjustments made to all of the tracks as the mix progresses. If after the guitars are added you find that the snare drum is not loud enough or needs to have a crisper sound, then these are the types of adjustments that are constantly done throughout the mixing process. These adjustments continue until the client and engineer are happy with the mix.

Creating multiple mixes

It is common to make different versions of a mix for the client to choose. These are generally different levels of the lead vocals. The lead vocals can be raised by 1 dB for a "vocal-up" mix. A "vocal-down" mix can also be created by lowering the overall level of the lead vocals. There can also be what is known as a "TV mix," which is a mix of all of the instruments with the exception of the lead vocals and perhaps the background vocals. This mix can be used for soundtracks or even to rerecord the lead vocals onto this mix in the future.

Mixing can be a very intensive process, so ear fatigue can set in depending on the listening levels. Always give yourself a break periodically to rest your ears. When you come back, you may hear different things in the mix that you did not notice before, or something that may have been bothering you before may sound fine after giving your ears a break.

Listening to the mixes

Take the time to bounce a mix to a CD and listen to it in different listening environments. Most people listen to music in their car, so many reference mixes are compared on a car stereo. If it seems that you are spending a lot of time on minute details in a mix, take a break from listening to that song for a day or two. Coming back with completely fresh ears and without the biases of having listened to the track repeatedly will allow you to hear things differently, and you can make adjustments according to what you hear after approaching the mix at a different time.

Remixing

Coming back to a mix to make adjustments is very common. Oftentimes, there may only need to be a few adjustments made to a few tracks. When approaching a new mix, make sure to save the remix as a different session so that you can always go back to the original if the remix does not wind up being any better than the original.

Archiving the project

Once all the mixing has been completed and the project has been mastered, the project needs to be archived in such a way that it can be remixed in the future. DAWs change over the course of time, so there is no guarantee that the project will play back as it is now in the distant future. Whatever DAW is common at that time may not be able to read the session file that points to where all the regions begin. Software instruments will certainly change over the course of time. Companies may go out of business, and there may be no way to re-create the exact sound used in a recording. This is where archiving the project so it has the best chance to be revisited in the future becomes important.

The best way to ensure compatibility in the future is to make sure that all of the audio files can be imported into any DAW. In order to ensure future compatibility, all of the tracks need to be audio tracks. The session should be saved separately so that you can leave the original final mixed file intact. It can be named something along the lines of "Song 1-Archive." To turn all of your tracks into audio files, begin by recording any software synthesizers that were used as audio files. These software synthesizers then become audio files that can be read in the future (Figure 7.22). This can be done by setting the output of the software synthesizer to zero, deactivating any plug-ins that were used, and setting the output to be an unused bus in the DAW. A new audio track can then be created with input set to the output bus of the software synthesizer. This track can then be recorded as an audio file.

FIGURE 7.22
String parts from a soft synthesizer being recorded as an audio file.

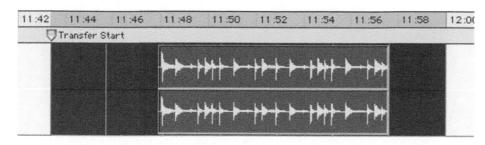

FIGURE 7.23
A selection of audio set to be consolidated from a single start point with the rest of the tracks.

FIGURE 7.24
Selecting the "Consolidate" menu in Pro Tools.

The same process should be done for any tracks that have Melodyne across them, as the audio coming from the Melodyne plug-in is being read from a separate audio file.

All of the audio files should then be exported in such a way that each one starts in the exact same location on the timeline. This can involve rewriting the audio file over the course of an audio selection. Some DAWs such as Logic will give you the option of automatically exporting all of the audio files so that they begin in the same spot for easy archiving and transferring.

If the song has a consistent tempo map, select a starting point for the audio files that falls directly on the downbeat of one of the measures. This will allow the files to be in sync in a different DAW by merely adjusting the tempo to the tempo of the original recording. In Pro Tools this can be done by selecting the beginning of a bar prior to any of the audio and placing a memory location at that point. This is easily done by pressing the "Enter" key. Then you can go track by track and rewrite the audio file from the start point to the end of the audio regions in that track (Figure 7.23). In Pro Tools this can be accomplished by highlighting a section to rewrite and going to the Edit menu and selecting "Consolidate" (Figure 7.24). The end result is a single audio file without any breaks in the regions (Figure 7.25).

Once all of the audio tracks to be consolidated have been selected, the audio files can all be highlighted and exported. DAWs will give the option as to what file type, sample rate, and bit depth to export the files as. There should be no

Edit	View	Track	Region	Event	Au
Undo Clear				⌘Z	
Redo Consolidate				⇧⌘Z	
Cut				⌘X	
Copy				⌘C	
Paste				⌘V	
Clear				⌘B	
Cut Special				▶	
Copy Special				▶	
Paste Special				▶	
Clear Special				▶	
Select All				⌘A	
Selection				▶	
Duplicate				⌘D	
Repeat...				⌥R	
Shift...				⌥H	
Insert Silence				⇧⌘E	
Snap to				▶	
Trim Region				▶	
Separate Region				▶	
Heal Separation				⌘H	
Strip Silence				⌘U	
Consolidate				⌥⇧3	
TCE Edit to Timeline Selection				⌥⇧U	
Automation				▶	
Fades				▶	

FIGURE 7.25
The final consolidated selection beginning at the same start point as the other tracks.

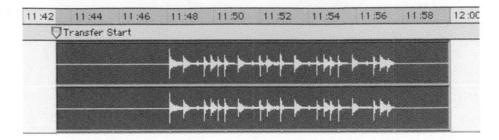

change as to the sample rate as this can create sound-quality loss inherent with any sample rate conversion. The file type should be the most commonly read file type by DAWs, which currently is the WAV file type. Export all of these files into a separate folder, and then transfer a copy of that folder to at least a couple of different types of media. DVDs can be easily read by computers, as can hard drives. CDs or DVDs tend to be the most universally read formats, and they can be read cross-platform. Hard drive formats can change over the years with the changing of operating systems and interfaces. Be sure there are at least two different copies of the archive. These archives should then be stored in two different locations away from the elements.

A FINAL WORD ON MIXING TECHNIQUES

Mixing a recording is a skill that takes years to develop. Being able to hear the different tracks and how the frequencies interact with each other takes time to develop. With today's DAWs it gets much easier to be able to go back to a mix and make revisions. It becomes easier to learn from your previous mistakes by being able to go back and analyze the way that you have previously mixed to a track. Remixing is a very common process, so do not hesitate to go back and make changes at the client's request. Sometimes it may help to scrap all of the previous work on a mix and start from scratch.

CHAPTER 8

Mastering the Recording

INTRODUCTION TO MASTERING

What is mastering?

Mastering is the last phase of the recording process. It is also the phase that is most often shrouded in mystery. Many inexperienced clients may not even know that mastering a recording is required. They may ask you why the rough mixes or mixes do not sound loud enough in the car. As an engineer, you will have to educate your clients as to what the mastering process entails. In summary, mastering takes your mixes, compresses, equalizes, and digitally limits the stereo two-track mix. The purpose of mastering is to make the mixes come alive when played back over ordinary speakers or on the radio.

If you are working on ten songs for a record, there will undoubtedly be tonal and dynamic differences between each of the songs. Mastering will balance out these subtle differences for each of the songs to create a sonically coherent record. The end result of mastering is the CD premaster, which will be sent off to the CD replicator.

Mastering can take place in either the analog domain, digital domain, or both. It depends on the capability and choice of the mastering engineer. Mastering equipment is different from ordinary studio equipment since it is designed specifically for mastering. Even if plug-ins are used, they are specifically designed for mastering as well.

If you are looking for detailed information about mastering, I would recommend checking out Bob Katz's *Mastering Audio—The Art and the Science*.

Steps of mastering

Mastering undergoes any number of steps and processes on the way to creating a CD premaster. There are no hard-and-fast rules when it comes to mastering—it always varies depending on the project. In any configuration, mastering can consist of any of the following steps:

- Equalization (linear and linear phase).
- Compression (full frequency and multiband).
- Digital limiting.
- Dither.
- Harmonic simulation (tube or tape emulation).
- Noise shaping.
- Stereo-field enhancement.
- Editing.

CD premaster

The CD premaster, which is sent off to the replicator, is assembled at the end of mastering. The mastering engineer will arrange the songs in the appropriate order, and then he or she will place any additional fades at the end or beginning of the songs that are necessary. He or she will then create the final premaster on a CD, which replaces the original format of the ¾-inch U-Matic tape.

Reasons for using a separate mastering engineer

Any mastering engineer will tell you that you should not master a recording yourself. There is some wisdom to this. Having somebody who has not lived with this recording for months can bring a fresh perspective as to what the final product should sound like.

Since mastering engineers can complete work on an entire record in one day, they will work on more projects a year than any recording or mixing engineer. A mastering engineer will know the current sounds of the day and create a master for you that is both musical and competitive to what is played on the radio.

An experienced mastering engineer will know how to use his or her tools inside and out. They know when to use them and when not to use them. You may think that you are saving yourself or the client a lot of money by mastering the recording yourself, but in the end you are paying more for the ears of the person mastering the recording than you are for the tools. Even the CDs that the mastering engineers utilize will be better, as there are certain types that are less prone to errors.

A professional mastering engineer's facility can have hundreds of thousands of dollars worth of equipment to allow him or her to select the appropriate tools for each project. The room will be perfectly tuned across the entire frequency spectrum, with monitors that will make any audiophile drool.

Preparing your mix for mastering

When you are mixing your recording, make sure that you are doing so in preparation for mastering. This includes making sure that you are not running anything across the stereo mix, including compressors, equalizers, or limiters. This kind of processing is added during the mastering phase. If you absolutely love a certain compressor that you have across the mix, give the mastering engineer two versions—one with the compressor and one without. He or she can then make a decision as to whether it contributes to the sound or not.

Mix levels for mastering

When you are mixing your song, always keep an eye on the stereo-mix levels; these are going to be the levels that the mastering engineer receives. If you see that you are peaking at or above zero, adjust the levels of the individual tracks accordingly. This can be observed in Pro Tools by pressing "Command" or "Alt" on your keyboard and simultaneously clicking on the level meter in Pro Tools. This will switch the number from reading the volume setting to showing the digital peak of the output. It is abbreviated by a "pk."

There is some debate as to what the best mix levels should be prior to mastering. The hard rule is to make sure the mix does not peak at or above 0 dB. The question then is where should the level peak? The safest way to go is to leave the mix some headroom below 0 dB. Having the mix peak between –6 dB and –3 dB allows some headroom for digital mastering with sacrificing little resolution. See Figure 8.1.

If the mastering engineer is going to use analog equipment for equalization and compression, the levels can be hotter than if everything is going to be done in the digital domain.

Sample rate and bit resolution for mastering

Almost every single digital audio workstation (DAW) today operates at 24 bits of resolution. The mixes that you give the mastering engineer should also be at 24 bits of resolution. This leaves more accuracy for the digital processing and the best-quality sound going through the mastering equipment, despite the fact that the final product will probably be 16 bits of resolution for a CD.

The files that you give the mastering engineer should be at the same sample rate as your mixing session. Do not do any type of sample rate conversion in order to bring your audio files to 44.1 kHz if the session was originally recorded at 48 kHz. The mastering engineer has better tools for sample rate conversion if he or she is working in the digital domain. If the mastering engineer is working in the

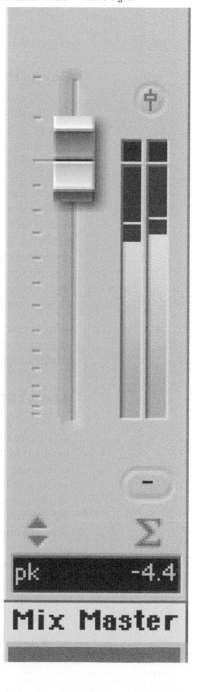

FIGURE 8.1
The mix-master fader of Pro Tools indicating the peak level of the digital audio signal.

analog domain, then he or she will resample the audio after it has gone through the mastering gear to the appropriate sample rate.

How much lead time and end time should the song have?

When you are creating your final mix, make sure that you have enough lead time at the beginning and end of the audio file. You can never really have too much extra time on either side.

Songs that fade out

Occasionally you may have a song where the artist wants the end to fade out. You can do this in a couple of different ways: You can create the fade in the mixed file, or you can create the fade in mastering. Usually it will be better to create a fade during the mastering stage, as that will produce the quietest results. If you are sending your mix out to be mastered elsewhere and you cannot be present during the mastering, then you have three different options. The first option is to give the mastering engineer a version of the mixes with the fade as you want it. The second option would be to give the mastering engineer a version of the audio without the fade and describe exactly where you want the fade to begin and end. The third and safest option is to give the mastering engineer a version of the mix without the fade, a version of the mix with the fade the way that you want it for reference, as well as a description of the fade. This allows the mastering engineer to create the fade the way you want it, but he or she can create the fade after all of their processing is done, which is ultimately cleaner.

When you are giving your mixes to the mastering engineer and you are looking for he or she to create the fades, make sure the audio in the mix goes on for much longer than you expect the duration of the fade to last.

MASTERING THE RECORDING YOURSELF

Now that we have discussed handing your project off to somebody else, it is time to go into mastering the recording yourself. There are many times that you may find yourself mastering a recording, even though you may have planned for a separate mastering session at a mastering house. Sometimes the client may need to have a version of the song to post on their web site or give to a publisher before the final mastering can be done. Sometimes the client has run out of money and needs to get the recording mastered before replication. You can still create a good-quality master yourself through careful listening and the utilization of quality tools.

Tools for mastering

If you plan on mastering a recording in your DAW, the best way to start is to gather the appropriate tools. Mastering tools should be of the highest possible quality. The main tools for mastering are:

- Equalizer (linear or linear phase).
- Compressor (single or multiband).

- Digital limiter.
- Redbook CD-authoring application.
- Dithering application (most limiters and DAWs will have these included).

There are also tools that additionally can be used for mastering that might include analog tape simulation, tube simulation, stereo spectrum enhancement, and metering. These can be handy tools that you should experiment with and have at your disposal.

To create a good-quality master you must first understand the standard signal flow for basic mastering.

Mastering signal flow

The basic flow of mastering is from the source to an equalizer, then to a compressor, then to a digital limiter, and finally to the destination audio file (Figure 8.2). Dithering and noise shaping are added at the very end, whether you are bouncing the tracks as a single file or in the CD-authoring program. The other optional processes can be added somewhere before the digital limiting.

This signal flow can be performed internally in a DAW or through outboard analog equipment. The only difference would be the D/A converter coming from the mastering source, and the A/D converter going into the mastering destination.

Setting up your DAW for mastering

Rather than try and perform mastering by placing plug-ins across the stereo-mix bus of the song that you are working on, it is best to create a whole new session with which to do your mastering. With a separate mastering session, it becomes easier to go back and make adjustments for each individual master. It is also easier to compare final versions of different songs so that you can make sure that all of your tracks sound sonically similar. The final reason for making a separate session for mastering is that you are using your computer's resources only for mastering and not mixing. High-quality mastering plug-ins can be much more processor intensive than plug-ins that are designed for mixing.

There are certain options that you will want to create when setting up your DAW for mastering. The first is that you will want to easily go back and forth between the processed and unprocessed versions; this can help you determine if your mastering is improving the audio in a manner that sounds the way that you want it to in the end. The second is that you will also want to be able to hear if the masters of the individual songs are sounding like a coherent record as well as having the same perceived loudness. Having all of the tracks in a single session will allow you to easily copy and paste plug-ins so that you can have a good starting point for each subsequent master.

FIGURE 8.2
A basic signal flow for the main processes of mastering.

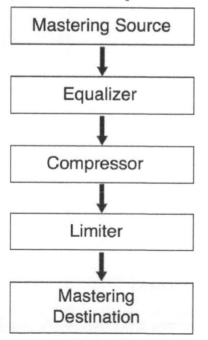

There are no set ways of accomplishing these goals, but a good way to set up your session for mastering is to import all of the songs into their own track. Then duplicate all of the tracks so that there are two versions of each song. Keep the original version as is. This will be the reference for the original unmastered version. Take the duplicate track and set its output to be Bus 1–2—this will be the track used for processing. Finally, create an empty stereo track after the duplicate track. Set its input to be Bus 1–2—this will be the destination for the final mastered track. Label the new destination track to be the mastered version of that song. You will now have the ability to solo and mute the different iterations of the songs to compare. See Figure 8.3 for an example.

With your session set up this way, you can place the processing across the duplicate track and then select input on the empty track. This will allow you to hear what you are doing with each step of the processing. After you have set up each of the plug-ins in the way that you want them, you are all set to record on the empty track.

When setting up the processing for the first track, you can compare the sound of the original unmastered version with the processing that you are doing on the duplicate track. This is where you can hear the difference of what the added

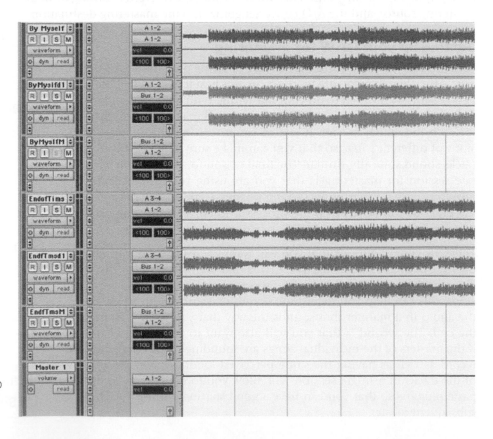

FIGURE 8.3

One of the ways that Pro Tools can be laid out for mastering.

processing is doing. Just go back and forth between the original track and the duplicate track and you have an easy way to compare the two versions, rather than activating and deactivating all the plug-ins.

After you have processed one track with your mastering plug-ins, you can move on to the next track. With your session laid out this way you can now compare the processing on the second track with the processing on the previously mastered

FIGURE 8.4
The mastered audio track recorded into Pro Tools. Notice the difference in overall gain.

track. This gives you the ability to compare the two and make sure that the sonic and dynamic qualities of both are comparable. See Figure 8.4 for an example.

When you have a final mastered version recorded on your empty track, you will need to edit the beginning and ending of the song. You can choose to place the fades at the end of these files; however, it is best to do that after you have processed all of the tracks. Some CD-authoring software packages give you the ability to add different fade curves to the songs, but at the end of this chapter there is a method of doing this all at once in your DAW. There may be a few instances in which you want the beginning of one song to start slightly into the fade of the previous song. This is an artistic choice, as it leaves you with maximum flexibility by saving it for the end.

Mastering plug-ins

There are many different plug-in options available for mastering. Since there are a variety of tools that can be used for mastering, these plug-ins are usually available individually or as a package of plug-ins designed for mastering. There are also plug-ins that are designed to incorporate all of the functions used in mastering, such as iZotope's Ozone. You may find yourself partial to one brand of equalizer and another brand of compressor, so having different options for each of these plug-ins can help you choose the right tool for the right job.

Digital equalization

There are many types of equalizers available in DAWs. Some equalizers function best as a mixing equalizer, while other equalizers are specifically designed for mastering. There are primarily two types of equalizers used in mastering. The first is a standard linear equalizer that functions similarly to an analog equalizer. The other is the linear phase equalizer, which is only available in the digital domain.

LINEAR EQUALIZATION

Most equalizers that are available in DAWs are linear equalizers, which have a similar sound to their analog counterparts. There is a big difference in quality from using a stock plug-in equalizer as opposed to using a higher-quality, high-resolution equalizer. Although there are some very high–quality linear digital

FIGURE 8.5
An McDSP linear equalizer dialed in for mastering.

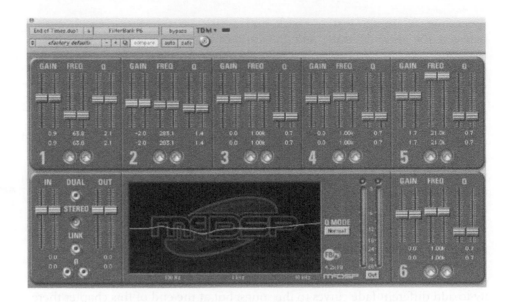

equalizers that are inexpensive, you will usually want to go beyond the stock equalizer that comes with most DAWs.

When you are mastering you may find that you need to make several small adjustments to the frequency spectrum. It becomes important to select a high-quality linear equalizer that will have several possible bands of equalization available; see, for example, Figure 8.5. Even though your equalizer may have six bands available, this does not mean that you have to use each one.

FIGURE 8.6
A Sonnox equalizer with the GML option.

Even though compressors and limiters are not designed to affect the frequencies of the audio running through them, they can have the effect of reducing some of the high-frequency content of the audio. Adding a slight high-frequency shelf with the equalizer can add more air to the track as well as compensate for the slight high-frequency loss that can be contributed from compressors and limiters.

If the particular equalizer that you are using does not have a shelving function, you can mimic a shelving equalizer with a parametric equalizer by bringing the frequency up as high as it will go and raising the gain of that frequency band. This does not have the exact same shape as a shelving equalizer, but using it this way can ramp up to the higher frequencies faster than a shelving equalizer. This may be preferable, and so experimentation with this equalization technique will give you another tool for your mastering toolbox.

Another thing that you may want to consider when equalizing tracks for mastering is to try to clarify the material. There may be some buildup of the low to mid frequencies and so lowering these slightly with the parametric equalizer will help clear up your recording in the mastering process.

If you are looking to add some more thump from the low frequencies, you can easily boost around 60 Hz with a high queue in order to bring out the feel of the kick drum. There are some generic equalization techniques you can apply, but each mixing engineer is different, and so the mastering needs for each project will vary.

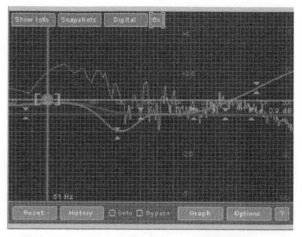

FIGURE 8.7
The iZotope linear phase equalizer. The curve appears more drastic, but the scale is set lower than other equalizers.

LINEAR PHASE EQUALIZATION

Linear phase equalization is almost exclusively designed for mastering. The filtering method used with linear phase equalization prevents any phase-related distortion that occurs with linear equalization. This is done with specific digital filters, either infinite impulse response filters (IIRs) or finite impulse response filters (FIRs). Essentially these linear phase equalizers are designed to eliminate any phase artifacts that may be created with a traditional linear equalizer.

Linear equalizers are designed to mimic their analog counterparts, and so they are mostly used as an insert for mixing. This is the sound that our ears have become accustomed to hearing across these tracks. Linear phase equalizers can have more dramatic "Q's", without the associated artifacts.

When using equalization it is important to A/B the equalizer on and off throughout the process. This will make sure that you are contributing to the sound of the final master. Most equalizers in a DAW will have the ability to turn on and off the different frequency bands. This works well when referencing to see if the specific equalization is working for the track. Once you have applied equalization to the necessary bands, A/B the entire equalizer to make sure that the end result sounds the way you want it to.

FIGURE 8.8
The Waves Linear Phase equalizer, which is part of Masters' bundle.

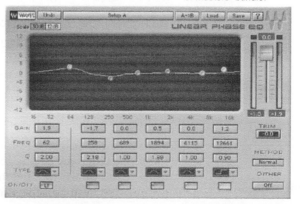

Mastering compression

Compression used in the mastering process is much different than the compression used during mixdown. The compression should be used to bring out subtleties in the music and slightly control the dynamic range of the material. Since the compression in mastering is usually much

FIGURE 8.9
An emulation of
the Fairchild 660
compressor.

more subtle than the mixdown process, the same compressor used during mixdown may not work for compression in mastering.

Many classic compressors are used more for the sound of their electronics, rather than the compression that they apply to the audio. Oftentimes these compressors are set up so that there is barely any gain reduction. There are many plug-ins that are designed to emulate these classic pieces of audio equipment (Figure 8.9). Experimenting with these emulated compressors can help you decide whether or not they are contributing to the quality of the master. Not all emulations are created equally, so you will need to use your ears.

Not every classic compressor is used in mastering. You would be better off selecting an emulated Fairchild over an emulated UREI 1176. This may require some research to find out which of the classic compressors that you have at your disposal would be useful for mastering and which ones would not.

There are two types of compressors used in mastering: the standard compression that affects everything across the track, and the multiband compressor that compresses different frequency ranges separately from the others.

DIGITAL SINGLE-BAND COMPRESSION

Using a standard compressor in mastering needs to be subtle like everything else in mastering. In general, you will want to use a low-compression ratio so that you do not hear the compression working heavily on the track. Select a low ratio and reduce the threshold with an eye on the total amount of gain reduction. The reduction in level can function effectively with as little as 1 dB or less of gain reduction. This is where you need to use your ears to determine what compression ratio and threshold work best for that particular compressor. Each compressor may sound different in what may be more noticeable compression than the other, so there are no hard-and-fast rules when it comes to compression in mastering, as long as there are no audibly distracting artifacts.

Using very low–compression ratios

Mastering compressors are designed to have much lower compression ratios than the standard compressors. When selecting a compressor for mastering, make sure that the compressor's ratio goes below 2:1. If the compressor's lowest ratio is too high, then the pumping of the compressor can become audible. Being able to go to a 1.2:1 or 1.1:1 ratio will help obtain the subtle compression you are looking for in mastering.

With these lower-compression ratios you can bring the threshold down substantially, and when the gain is compensated correctly, you hear the subtleties of the music come up without hearing the upper end of the dynamic range being compressed downward. See Figure 8.10 for an example.

In general, the speed of the compressor should be set for a fast attack and slower release. The gain reduction should be subtle—anything more than 3 dB and you may start to get obviously noticeable pumping of the audio.

MULTIBAND COMPRESSION

Multiband compression is one of the options available for controlling the dynamic range. This method of compression separates out the frequency spectrum into various bands; see Figure 8.11 for an example. The frequency crossover point of these bands can be adjustable. Multiband compression can be tricky to use effectively. It may take some trial and error before you get quality results. Compressing different ranges of the frequency spectrum has the advantage of being able to compress the low-frequency range to add some consistency to the low end. You can also add some subtle compression across the mid- and high-frequency regions.

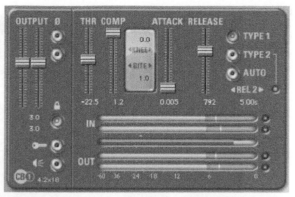

FIGURE 8.10
An McDSP equalizer set for a 1.2:1 ratio with a low threshold.

Compressing across the different frequency bands essentially functions as having multiple compressors. Each of these compressors can have their own threshold, compression ratio, attack and release times, and makeup gain. This means you can compress the low frequencies with a higher ratio than you can with the rest of the audio.

If you are new to mastering, multiband compression may not be the compression type that is best suited for your first couple of projects. You can easily do more harm than good using multiband compression. To start out, just try compressing the low end of the spectrum using a multiband compressor and then using a single-band compressor for overall compression.

PARALLEL COMPRESSION

Parallel compression can be used in mastering much in the same way it is used during mixdown. This can be helpful to contain the overall dynamic range of the different sections of a song. If there are soft sections that are coming across as being too soft in the final master, parallel compression can help raise the level of the soft sections without making the loud sections sound too compressed.

FIGURE 8.11
The Waves multiband compressor, utilizing five bands of compression.

The session setup when using parallel compression is slightly different than the standard mastering setup in that you have to duplicate the track. Depending on the system that you are using, there may be latency involved, which can create phase cancellation. In order to avoid this, you can merely copy the same compressor plug-in from the compressed track to the uncompressed

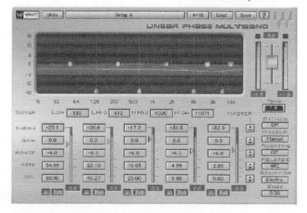

FIGURE 8.12
The McDSP MC2000 multiband compressor with four bands of compression.

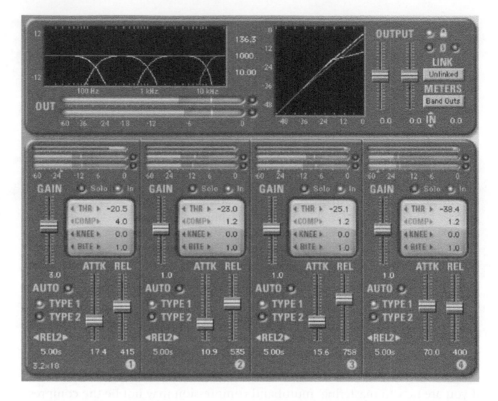

track. For the uncompressed track you can then place the compressor plug-in in bypass. This will create the same latency that you have with the compressed track (Figure 8.13). Make sure that you copy the identical plug-in setting for any processing you used prior to the compression; otherwise you will generate phase-cancellation artifacts.

FIGURE 8.13
Parallel compression being utilized in mastering. The upper track has the compressor placed in bypass to maintain a consistent latency.

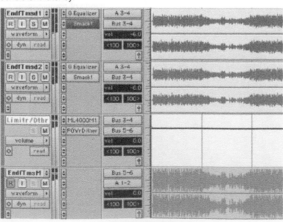

These tracks can be combined into a single recorded track. If you are looking to add a digital limiter across the parallel-compressed audio tracks, you can create a separate auxiliary track after both of the compressed and uncompressed tracks but before the final recording master. This allows you to apply a limiter and dither, but still make adjustments in real time before you record your final mastered track.

In Figure 8.14, you can see the end result of the parallel compression. The duplicated tracks have been recorded onto the third track. With this track, you can see that the dynamics of the louder section remain consistent, while the softer section has been brought up in level with the parallel compression.

Digital limiting

Using a digital limiter as the final stage of signal processing during the mastering process has
become common practice for the past several
years. Digital limiters give recordings their perceived loudness. For the past 15 years CDs have
gotten progressively louder and louder through
the use of these limiters. The end result is that
many of these records have lost the subtlety of
their dynamics.

FIGURE 8.14
Parallel compression
recorded onto a track in
Pro Tools.

If you just want to make a track sound loud,
then a digital limiter will easily do the trick. Oftentimes, mastering is associated with making a record louder. This is usually the shortest step of the mastering process, as the real mastering takes place with equalizing and compressing
of the stereo mixes through the careful listening of the professional mastering
engineer.

Limiting takes place in the last stage of the mastering process, with the exception of any dithering or noise shaping you choose to add. Some digital limiters
give you the choice of release times with the attack times functioning with the
plug-in's own proprietary algorithm. These digital limiters look ahead of the
audio to see where the peaks are and then reduce the gain with their own proprietary algorithms in such a way that there is no clipping of the audio signal
while increasing the gain.

Lowering the threshold of these digital limiters will simultaneously increase the
makeup gain. Therefore, if the digital limiter's threshold is reduced by 6 dB, the
digital limiter automatically applies 6 dB of makeup gain. This will instantly
make the perceived loudness of your master increase the moment you start
decreasing the threshold.

FIGURE 8.15
The McDSP digital
limiter.

There is a certain point when digital limiting will turn into distortion. Depending on how much compression you put across a track,
this will determine how soon you will hear the distortion that is
added by the digital limiter. For a track with very little compression, you can get away with 6 or 7 dB of gain reduction. If there is a
moderate amount of compression across the master track then 2–4
dB of gain reduction will be sufficient.

A handy trick to hear the colorization that is being applied by
the digital limiter is to reduce the ceiling equal to the reduction
of the threshold. If the reduced threshold is adding 6 dB of gain
reduction, and the ceiling is being reduced by 6 dB, you will hear
the limited version of the track at the same perceived level as the
unlimited version. From here, you can put the digital limiter in and
out of bypass to hear the colorization or distortion as the digital
limiter is adding to the master.

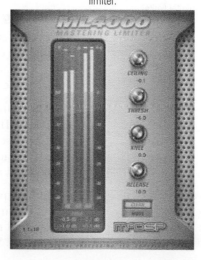

The other parameter in digital limiters that you will need to adjust is the ceiling. The ceiling determines the maximum level of decibels that will be output from the digital audio file. Since digital audio files max out at 0 dB, the ceiling becomes a number lower than zero.

Digital audio playback devices such as a DAW or CD player have a certain point at which they detect clipping of the digital audio. Since digital audio does not go above 0 dB, these devices detect clipping by a specific number of consecutive samples at 0 dB. The number of consecutive samples varies depending on the manufacturer. Reducing the ceiling so that it is slightly below 0 dB will prevent these digital audio devices from displaying any audio clipping. There is no specific standard where the ceiling should be lowered. Contemporary digital audio recordings will have ceilings anywhere from –1 dB to –0.1 dB, with –0.5 dB and –0.1 dB being the most common.

Distortion can potentially occur in the digital-to-analog conversion of the end user's playback device, such as a CD player. Slightly lowering the ceiling of a digital limiter will help prevent this distortion.

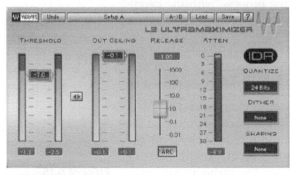

FIGURE 8.16
The popular Waves L2
digital limiter.

There are many different digital limiters available on the market. Waves' L2 has been the most common, as it has been around the longest (Figure 8.16). In the past few years, there have been competing digital mastering plug-ins available on the market. These include McDSP's ML4000, Massey's L2007, and Voxengo's Elephant, to name a few. Prices vary tremendously for each of these, from $69 all the way up to $1200. It depends on the level of functionality you are looking for, but professional mastering engineers have used each of these digital limiters.

Dithering, noise shaping, and quantization

Since most current DAWs function at 24 bits and most often the end product is 16 bits (the CD standard), there needs to be some processing to reduce the 24-bit audio in quantize to 16 bits. This quantization requires specific signal processing to reduce noise generated by quantization error. This process is known as dithering.

Dithering and noise shaping occur simultaneously with the final quantization reduction. The levels will differ to the bit depth being quantized, so choose your final bit resolution in your dithering plug-in. Even if you are recording 24-bit files, if the audio is dithered to 16 bits and quantized as such, the last 8 bits will all be zeros, so truncating these will not make a difference.

Dithering and noise shaping are the very last pieces of signal processing added during the mastering process. Since digital limiting winds up being used at the last stage of mastering, some digital limiters will have dithering and noise shaping built in. Waves' L2 is an example of this. If your digital limiter does not have this function, DAWs will have at least one plug-in that can create dither and also noise shaping.

The science behind dithering and noise shaping is too complicated to explain in a chapter such as this. The best way to describe dithering is that it reduces noise added by quantization error at low levels. It accomplishes this by adding a low-level noise that masks the distortion. Noise shaping works with dithering and is designed to increase the perceived dynamic range.

For most dithering, many DAWs will include the POW-R (Psychoacoustically Optimized Wordlength Reduction) dithering application (Figure 8.17). There are three different algorithms associated with POW-R, and each one is designed to work with a specific type of program material. The number of the algorithm correlates to the increasing complexity of the program material. The first is designed to work best with solo instruments. The second is designed to work best with pop music, and the third is designed to work best with orchestral music.

FIGURE 8.17
The POW-R dithering plug-in, quantizing the audio and adding noise shaping.

Dithering should always be the very last step of the mastering process, since any changes made to the audio have an effect on the distortion-reducing effects of dithering. These include any fades or crossfades you may apply. If you plan on adding fades or crossfades to the different mastered tracks once they have been processed, plan to add the dithering after all of these other processes. Since this is the last step of the mastering process, all that is left to do is to record your final mastered tracks in your DAW.

WHEN TO APPLY DITHERING AND NOISE SHAPING

These processes should be applied at the very last stage of mastering. You can still equalize, compress, and limit your tracks for mastering without necessarily applying dithering and noise shaping at the same time. If you are looking to sequence the music in such a way as to apply crossfades or overlaps, it is best to wait and apply dithering, noise shaping, and quantization after you have made all the edits and are making the final export for your CD-authoring program.

When exporting a mix to 16 bits, DAWs are defaulted to apply a dithering to the exported audio file. If you are applying dithering through the use of a separate plug-in, the dithering noise added will be cumulative, so you will want to turn the extra dithering off in the preferences of your DAW.

Checking levels and metering during mastering

Since you are looking to create a coherent record, if you are mastering several songs, you need to check the levels of each track throughout the course of mastering to hear if the perceived volume is the same for each track. There are different metering tools available to check the levels during mastering; however, there should be more concern as to how the songs sound with each other as opposed to how the meter looks.

As you are mastering your tracks, keep an eye out on the digital meter to make sure there is no clipping between the different plug-in instances or the stereo master meter. Clipping can occur at the output of plug-ins, so you may need to adjust your output gain of the offending plug-in if this occurs.

Just looking at the peak metering will be no help once the audio has gone through a digital limiter. This is because the peak is now determined by the ceiling of the digital limiter, which should be the same for each track. In order to use a meter to check the level of the final result, you will need to find a metering plug-in that will allow you to examine the levels with RMS (root, mean, square) metering. This is an averaging meter that will tell you what the average level is.

FIGURE 8.18

Two meters comparing the RMS level of different mastered tracks.

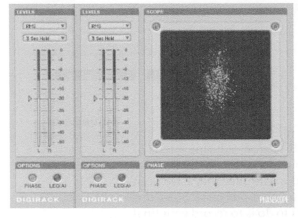

If you are mastering an entire CD or EP, you may find that certain songs may not have the same RMS readings as the others. You may have a slow song that may seem extremely loud if it is set to be perceived at the same loudness as the faster tracks.

If you see an average level difference, then you may hear an audible difference in the perceived loudness of each mastered track. You can make adjustments to the perceived loudness by simply adjusting the threshold of the digital limiter up or down. If there is a question as to whether you should lower the threshold on the softer track or raise the threshold on the louder track, it may be best to raise the threshold on the louder track to bring its perceived loudness down. This will help make the music more dynamic as well as preventing distortion contributed by the digital limiter.

FIGURE 8.19

Ozone's real-time analyzer displaying the frequency content of the source material.

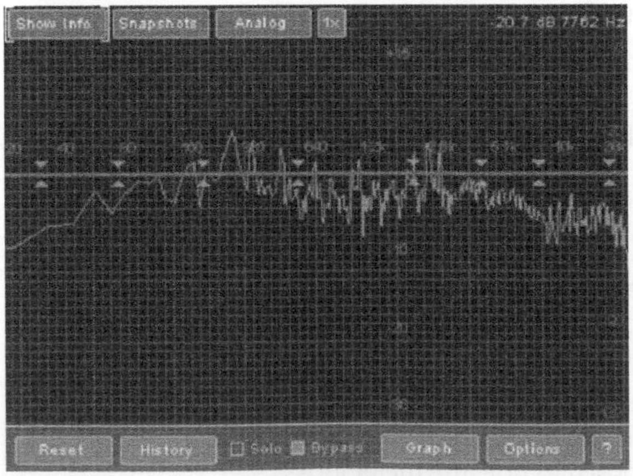

In Figure 8.18, you can see that the RMS level of these two mastered tracks is different. Since they are of the same style of music, there is an audible difference in the perceived loudness of each track.

You can also use a real-time analyzer, which will give you the levels across the frequency spectrum. You can use this to detect whether there is a specific frequency buildup or deficiency. If your monitors are not capable of playing back low frequencies, you can use the real-time analyzer to check the content of that frequency range. The final master should not need to look flat across a real-time analyzer, so do not become too fixated on what it looks like; see, for example, Figure 8.19. The more experience you have with a real-time analyzer, the more effective of a mastering engineer you can be by using one.

Other mastering processes

There are many other tools that can be used in mastering beyond just equalization, compression, and limiting. There are also tools that allow for adjustments to the stereo spectrum and analog simulation. These tools are certainly optional and should only be used subtly and for specific needs.

ANALOG SIMULATION

There are many different analog simulators on the market. These are designed to mimic specific qualities of analog tape. Analog tape can create a smoothness to the sound as well as some compression and a variation to the frequency response.

One of the properties of using analog tape is that it creates compression to the recorded material. If you are using an analog tape simulator in your mastering chain, keep this in mind if you are going to add separate compression on top of analog tape simulation. Some analog simulators, such as McDSP's Analog Channel (Figure 8.20), will display the amount of gain reduction being applied to signal.

There is also an equalization that applies when you are using analog tape, and these analog tape simulators will incorporate these frequency changes. The first most notable frequency change will be in the low frequencies. There will be a specific bump around 100 Hz or below. This varies with the speed of the tape machine. The other frequency change will be in the high frequencies, around 10 kHz. The high-frequency adjustment generally deals specifically with the biasing of the tape recorder being emulated. Some analog simulators will allow you to adjust the biasing so that you can tailor the high-frequency adjustment to suit your needs.

The placement of an analog simulator in your mastering signal chain can vary depending on what you are trying to do with your processing. Since many recordings are mastered from a ½-inch analog tape, mimicking this would be

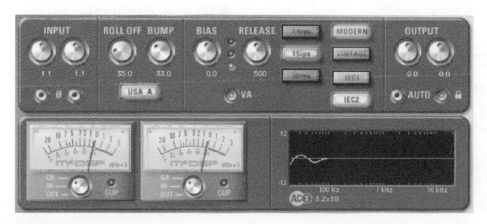

FIGURE 8.20
The McDSP Analog
Channel 2 tape emulator.

FIGURE 8.21
The Cranesong Phoenix adds harmonic content and compression to a track.

to place the analog simulation at the very beginning of the signal chain. This would be as if you were mastering straight off the analog recording. If you are looking to utilize the analog simulator as a compressor, place the plug-in after the equalization. If you want to make the end result sound as if it was coming off analog tape, then place the analog simulator just before the digital limiter and dial in the settings to make it sound as close to the analog tape as possible. This will work best with a more neutral analog simulation as opposed to more colored analog sound.

SPATIAL ENHANCEMENT

There is a specific type of signal processing used in mastering to enhance the stereo field. This gives the mastering engineer the ability to narrow or widen the stereo spectrum. There are many different algorithms available to carry out spatial enhancement. These can involve increasing or decreasing the gain of material that can be canceled out from each side. Another method can involve widening the sound based on frequency.

Widening the stereo field can be tricky and can also sound artificial. Low frequencies are usually localized by the listener from every direction. The higher the frequency, the more we can perceive directionality. Trying to widen the stereo spectrum of low frequencies will not have near the effect as it will to high frequencies. Therefore, if any spatial enhancement is to be done to the audio it may be best to apply them at the very high frequencies where they will be more natural.

FIGURE 8.22
Ozone's multiband stereo-imaging enhancer adjusting the width of the different frequency bands.

In addition to widening the stereo field, spatial enhancement can be used to narrow the stereo field. This can be used across low frequencies to narrow them in the center, and it can be a great tool to focus delivery to the center, leaving the higher frequencies more room in the stereo spectrum.

iZotope Ozone

iZotope makes a high-quality mastering package that functions as a single plug-in instance. It contains most of the tools that you can possibly ask for when mastering. This becomes a great solution for the budget-conscious engineer who is looking for some high-quality tools for mastering. It includes an equalizer, mastering reverb, harmonic exciter, multiband compression, stereo enhancement, and a digital limiter. Despite being a single plug-in instance you have flexibility with the order in which the processes are applied, as well as numerous options with each process.

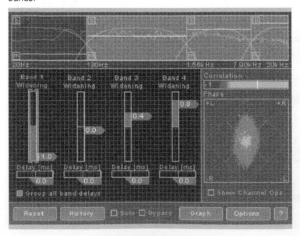

Each module in iZotope's Ozone is defaulted to be turned off. As you begin each module, start by clicking each one and making it active. You also have the ability to solo a specific module so that you can hear that module's direct effect on the overall master. You can then turn the bypass on and off of that module to compare that module's effect to the original unmastered recording. Just because there are six different modules available, do not feel as though you need to use every single one.

OZONE EQUALIZATION

The equalizer in iZotope's Ozone is powerful, yet easy to use. There are two modes of equalization. The first is the standard linear equalizer, which is represented in Ozone as being analog. The other is a linear phase equalizer, which is represented in Ozone as digital. Switching back and forth between these two modes is simply a matter of clicking on the analog or digital button to switch to the other. There is also a matching mode to the equalizer, which we will get into later. The equalizer contains six nodes—a parametric equalization, low and high shelving, and high- and low-pass filters—as well as two additional nodes that can be either filters or shelving equalizers.

Clicking on each node of the equalizer will allow you to boost/cut and adjust the frequency. When you click on each node a bracket forms around the selected node, which allows you to widen or narrow the Q of the equalizer. As you play back the audio through the plug-in, you can see the real-time analyzer displayed in the background. This can give you an idea as to where the low frequencies of the kick drum hits are at, as well as help you detect any low- to mid-frequency buildup.

For mastering, hopefully there should be no need to do any large boosting or cutting of frequencies. Zooming in on the vertical axis of the equalizer will help you make small adjustments so that you can see them visually on the screen.

MATCHING EQUALIZER

One of the more interesting and unique features of iZotope's Ozone is the ability to match the equalization of a previously recorded commercial recording. This feature analyzes the frequency spectrum of another recording and compares it to the frequency spectrum of the track that you are mastering. It then creates an equalization curve that can help make your recording match closely with the sonic spectrum of the commercial recording. This can be helpful if you are trying to achieve a specific tonality for your recording matching that of a commercial recording.

To begin creating a matched equalization, select a song that has a similar style and tonality that you are looking for to match your current track to. Import this song into your DAW and place it in the same audio track as the song that you

FIGURE 8.23
The analog emulated equalizer in Ozone.

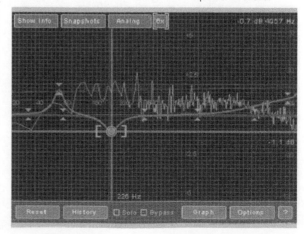

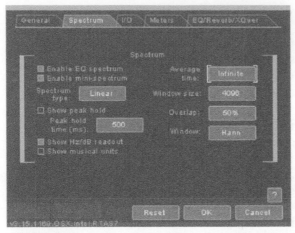

FIGURE 8.24

Ozone's settings for the matching equalizer.

FIGURE 8.25

The captured frequency response of the source material.

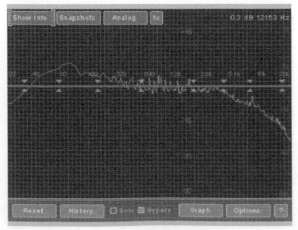

are currently mastering, just after the song in the timeline. Make sure that your song and the commercial recording share a similar overall tonality; otherwise, the matched equalization may be more dramatic than you are looking for.

To create a matching equalizer you need to set the Ozone's equalizer to the appropriate settings (Figure 8.24). The matching equalizer works separately from the analog or digital modes. Move the setting to "Matching" after you have taken the appropriate snapshots of the frequency spectrums.

Now you are going to take snapshots of both the source track and the destination track in order to create a matching equalization curve. You will need to take the analysis of the spectrum and set the average time to infinite. This will continuously adjust the analyzed frequency spectrum based on the material coming through the real-time analyzer of iZotope's Ozone. This is found under the Options menu by clicking on the "Spectrum" tab.

Next you will need to make a snapshot of the frequency spectrum coming in to iZotope's Ozone. You can begin the snapshot process by clicking in the middle of the analysis window to reset the real-time analyzer, and then begin playing the source track to capture the frequency spectrum (Figure 8.25).

Play back a selection of the song that consists of the main frequency content that you are looking to capture. You will see the analyzer window adjust the curve over the course of time and then settle into an analysis curve that does not change very much. You should not need to play back too much of the selection; usually only 10–15 seconds will be sufficient. While the track is still playing, but it has settled into a consistent analysis curve, click on the "Snapshots" box. Then click the letter and color you want to be your source, and click "Source." This will create a snapshot for that analysis curve.

After you have taken a snapshot of the source analysis curve you will need to capture the analysis curve of the destination material. This is the same process, but be sure to click in the analysis window to reset the real-time analyzer. Play back a 10- to 15-second chunk of the destination material and click on a different letter. Select that letter as being the target equalizer curve by clicking on the "Target" button. You will probably see a difference in overall levels between the two frequency spectrums, but the matching equalizer will not adjust the overall level for the frequency content, so this should not be an issue. See Figure 8.26.

After you have the two snapshots of the frequency spectrums, you can now switch the equalization mode to matching. Ozone now can create an equalization curve based on the differences of the sonic spectrum between the two snapshots. Inside of the snapshots window there are two adjustments that you can make for the matching equalizer. The first is the percent of matching amount, which will be the amount of the equalization difference from the matching equalizer you wish to bring in to the destination material. The second option to adjust is the smoothing. Since the matching equalizer of iZotope's Ozone utilizes 4096 bands of equalization, this can create a very jagged equalization adjustment curve. The smoothing feature will smooth out the jagged edges of the curve and make for a more realistic and natural-sounding equalization curve. Adjusting the matching amount to be 50 percent and the smoothing amount to be at 0.5, which is in the center, will give you a much more natural starting place. From here you can experiment with how much matching and smoothing you want to do with your matching equalizer.

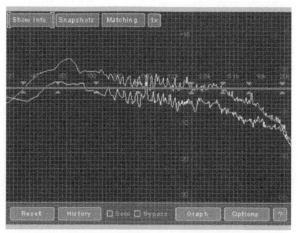

FIGURE 8.26
The two different equalizers snapshots captured.

In Figure 8.27, the matching amount has been set to 100 percent, and the smoothing at 1 (maximum), while in Figure 8.28, the matching amount has been reduced to 50 percent and the smoothing set at 0.5.

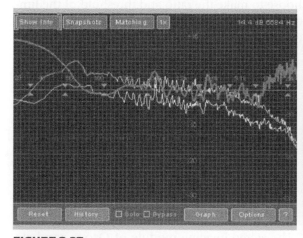

FIGURE 8.27
The two equalization snapshots with the compensated equalizer added.

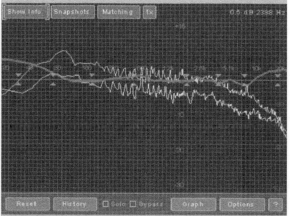

FIGURE 8.28
The matching equalizer with the compensated equalizer smoothed.

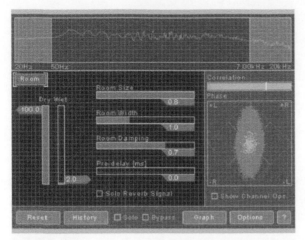

FIGURE 8.29
The global reverb, blending 2 percent of the wet signal into the master.

MASTERING REVERB

Ozone gives you the option of adding a reverb to your mastered track. This is not a common process but can be used to add a little bit of space to the master. If you choose to add any additional reverb to your master, realize that it is going across the entire track as opposed to reverb in mixing. Using a tiny bit of the mastering reverb can potentially add some smoothness and life to your recording. See Figure 8.29 as an example.

MULTIBAND DYNAMICS

The multiband dynamics module of the Ozone functions in the same way as other multiband compressors. It features a compressor, expander, and limiter on each of the four frequency bands. It is important to note that the multiband dynamics, multiband harmonic exciter, and multiband stereo imaging all share the same crossover points with each other. If you adjust one crossover point in the dynamics section, that same crossover point will be adjusted in the stereo-imaging section. This makes it important to adjust your frequency bands during one of these processes and stick with them, or make only slight adjustments throughout the mastering process.

In addition to featuring a compressor across each of the frequency bands, there is also a limiter with an adjustable ratio. Rather than using this as a limiter you can reduce the ratio as low as the ratio for the compressor. This will allow you to use this as two separate compressors rather than a compressor and limiter. Using two compressors in this fashion can give the effect of having a softer knee on the compressor. You can choose to make the ratio and threshold of the limiter slightly above that of the compressor. This will then create two compressors functioning in a series (Figure 8.30). There is no individual gain for the different frequency bands, so you will have to create any makeup gain with the global gain of the module.

FIGURE 8.30
The limiter being added as an additional serial multiband compressor.

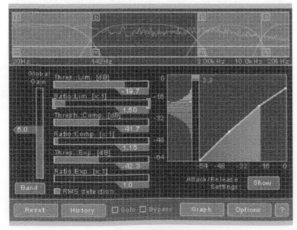

When making many different adjustments involving the attack and release time, compression ratios, and thresholds, you can copy and paste any of these settings by right-clicking on the specific parameter and then moving and pasting them onto the next one.

STEREO IMAGING

The stereo-imaging module of Ozone can help enhance the spatialization of the different frequency bands. You have the ability to widen or narrow the stereo spectrum across each of the four different bands. You can mute and bypass any of the different frequency bands from what you have done in the multiband stereo imaging by clicking on the "B" or "M" on each of the crossover points at the top of the window. This will help you hear the difference with the adjustments that you are doing on each band. See Figure 8.31.

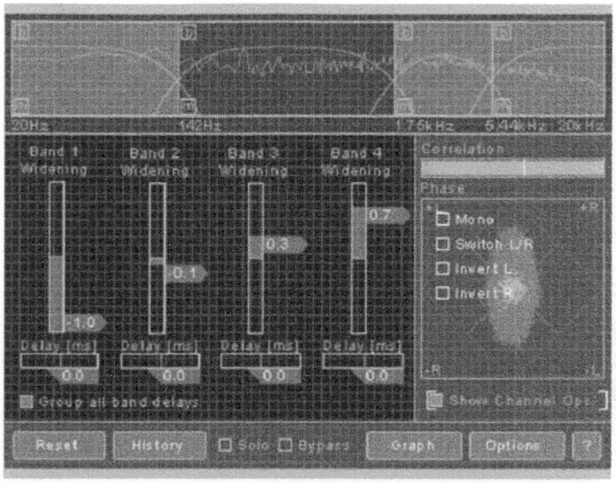

This processing will not create any stereo information that is not already in stereo in the unmastered track. It will only enhance the width of the audio on each of the frequency bands.

You can select "show channel ops" at the bottom of the oscilloscope, which will allow you to click "Mono" to check the mono compatibility of your track. This can help you determine how much widening and narrowing you want to do with each frequency band. As with any stereo enhancement keep in mind that the low frequencies are far less directional than the higher frequencies.

FIGURE 8.31
The multiband stereo-imaging module.

MULTIBAND HARMONIC EXCITER

Ozone also contains a multiband harmonic exciter (Figure 8.32). There are many different types of harmonic exciters on the market that create additional harmonic content derived from the audio material. Essentially, in Ozone, this functions as an analog tape simulator or a tube simulator. Each one of these simulations has a different sonic quality, so experiment with different types, as they each will create different harmonic content. You have the option as to the amount of harmonic content created as well as the blend between the harmonically excited audio versus the dry signal. This is another one of those additional options in mastering that can create some subtle enhancement if used appropriately. A little bit of the multiband harmonic exciter can go a long way.

FIGURE 8.32
Ozone's multiband harmonic exciter.

LOUDNESS MAXIMIZER

Ozone's loudness maximizer is its digital limiter. The threshold and margin function in the same way as they do in any other digital limiter. There are also different settings for the type of limiting.

There are three different modes for the limiter. First, there is a soft mode, which will allow transients to go beyond the margin depending on the material. This is a more natural-sounding limiter. If this is the mode that you want to use, carefully

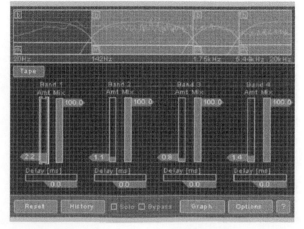

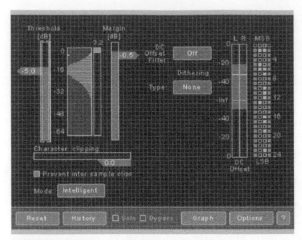

FIGURE 8.33
Ozone's digital limiter set to intelligent mode.

FIGURE 8.34
Ozone's digital limiting module with added dithering and noise shaping.

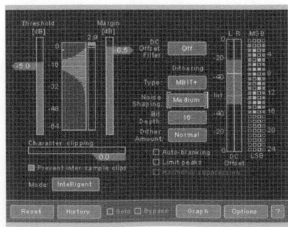

monitor the output to make sure that no digital clipping takes place. Second, there is a brickwall mode, which will not allow any transients beyond the margin. For most pop-record mastering, the intelligent mode will be the best choice. Finally, the intelligent mode functions the same way that many of the other digital limiters on the market work. It has an adaptive release time and analyzes the transients before it applies the limiting to make for a more transparent limiter. See Figure 8.33.

There is a character slider, which allows you to adjust the release time of the digital limiter. The slider will only work if you are utilizing either the soft or brick-wall modes of the digital limiter, not the intelligent mode. You will see that the adjectives following the character describe what the release time will do to the audio. These range from clipping, very fast, fast and loud, smooth, transparent, and very slow. Decide which character is right for the material.

DITHERING

The dithering and iZotope's Ozone is located in the same module as the loudness maximizer. iZotope has its own proprietary dithering, which is referred to as MBIT+. There are also two other types of dithering referred to as Type 1 and Type 2. The MBIT+ works as the best default dithering in Ozone. It comes with its own noise shaping with a selectable amount from none to ultra. You also have the ability to select how much dithering you wish to apply. These range from low to high. See Figure 8.34.

One interesting thing that you will notice is the effect of dithering and noise shaping on the digital audio output. Once these are engaged you can see that dithering and noise shaping raise the noise floor when they are applied. With MBIT+ dithering and ultra noise shaping there is approximately 35 dB of noise added to the end result. You can see how dithering multiple times can eventually create an audible noise in the end product (Figure 8.35).

Assembling the tracks for your CD-authoring program

If you are mastering more than one track and looking to create a CD premaster for replication, you need to sequence the songs in the appropriate order and place any desired crossfades in between the songs. If you are mastering a single song, then merely exporting the final master with the appropriate dithering and noise shaping is sufficient for most purposes.

Many of the CD-burning programs on the market allow you to create a CD premaster. These products have the ability to make adjustments to the level and apply crossfades and perhaps even some digital signal processing. These are all processes best done in your DAW or CD-mastering software. The main feature of the software programs that you will use is the ability to adjust the timing between tracks.

Once you have processed all of the songs, you can then drag them into a new session in your DAW. It is best to place each mastered audio file into its own track or be able to alternate between two tracks for overlapping. Drag each of the processed audio files into the appropriate order in the timeline so that you can play through the CD in sequence. You can trim the start and ending of each song in case they are longer than what you want for the final master.

FIGURE 8.35
The noise generated by noise shaping is visible on the output of the digital meter.

If you are looking to create a tight distance between songs there may be some slight overlap as one song fades out while the next song can be set to begin at the tail end of the fade. You can place a very slight fade at the beginning of the audio file. See Figure 8.36.

Once you have all the songs sequenced and played through them to hear how the entire mastered recording will sound on the final CD premaster, you can now bounce the entire track as a single stereo audio file (Figure 8.37). Make sure that you have applied the dithering and noise shaping across the master bus. Now that you have a single stereo audio file you can edit each

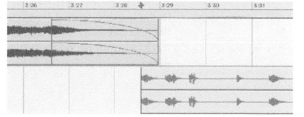

FIGURE 8.36
Two mastered tracks with the first one fading into the next.

of these tracks to place the appropriate ID number for each track. This can be done in a completely separate session in your DAW.

After the track has been bounced, place markers in the session to denote the start of each track. You can then jump to the start of each track to make sure that you are not cutting off the start of the song and that the ID number will not have

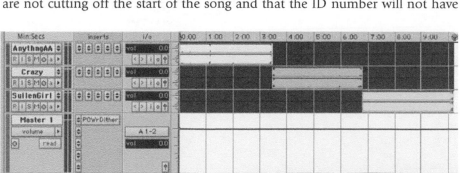

FIGURE 8.37
Three tracks being bounced as a single stereo audio file.

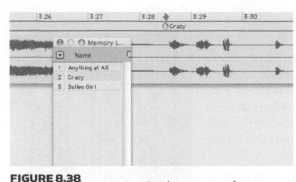

FIGURE 8.38
Memory locations placed along the composite stereo master to denote the start time for each track.

too much of the tail of the previous song playing. Place a break in the audio file at the beginning of each marker. Now you can separate the regions inside your DAW. Each of these audio files becomes the corresponding CD tracks. Label each region as the song name.

You can now export all of these regions as individual audio files. These should be in the final format of 16 bit, 44.1 kHz. It can be handy to put the song number at the beginning of the region's name so that you can easily place them in the appropriate order in your CD-authoring software. Rename these files beginning with 01, 02, 03, etc. If you are using Pro Tools as your DAW, make sure the files are exported as stereo interleaved files, as opposed to (multiple) mono files.

Once the audio has been imported into your CD-authoring software, make sure that they appear in the right order. Using Roxio's toast or Easy Media Creator, you will need to change some of the default settings. Most CD-burning programs will automatically have the dither set to be applied to the audio. Since we have already dithered the tracks, switch this off in the preferences.

There is a pause setting that defaults to a two-second pause between tracks. Since we have already sequenced our regions so that they are supposed to play next to each other, adjust this pause to zero seconds (Figure 8.39). The first track on the CD needs to have a default of two seconds, but the pause between the other tracks should be set to zero seconds. There are also default crossfades listed, but these are not active, as there has been no crossfade duration set.

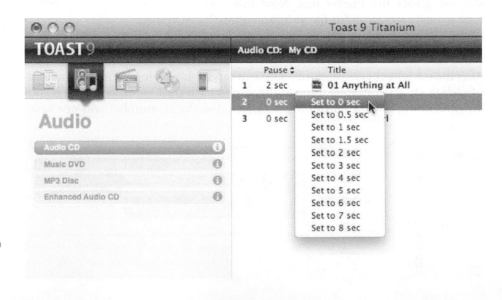

FIGURE 8.39
Setting the pause between tracks to zero in the CD-authoring program.

With the tracks in order, play through each one and rewind before the beginning of each ID number to make sure that each track transitions correctly. You can now burn your CD premaster. Make sure that your recording program is set to record the "Disc at Once" (DAO).

Select a high-quality CD for your premaster. There will be a difference in the number of errors on a high-quality CD as opposed to a cheap, generic CD. There are many forums where people recommend various types of CDs for mastering. Among the most popular are Taiyo Yuden Silver or Green Tune. The next choice to make is the speed at which to burn your CD premaster. The faster that a CD recorder is set to record, the more errors that it may generate on the CD. Most current CD recorders will not record at the 1× or 2× speed that many recommend. The best thing to do would be to choose a slower speed and then listen to the CD for errors.

Checking the CD master

Always play through the entire CD before you send it off to the replication facility. There can always be problems, no matter how expensive a CD you have chosen. You can use less-expensive CDs to give to the client as a reference.

A FINAL WORD ON MASTERING THE RECORDING

Mastering, like mixing, can require some updates to the work done. When mastering on a DAW, it makes it easier to go back and make subtle adjustments to the equalization, compression, or limiting done on a record. Even if you have done some mastering yourself, you may find that taking the mixes into a professional mastering facility can be an eye-opening experience. With all types of engineering, it is advantageous to be able to observe and learn from other professionals.

Conclusion: A Final Word

The music industry has always been constantly changing. This includes the recording and production techniques used to create these recordings. This book has presented many of the current techniques used by professionals. It is by no means a comprehensive listing of the tools used, as it would be impossible to fit them all in a single book. The techniques presented, however, can be applied to different tools that are used to achieve the same goals, such as pitch correction and sound triggering.

The world of musical synthesizers expands every year, with more and more instruments being available in software as opposed to hardware. Now that these devices are inside the computer, musicians and engineers need to know the practice and function of these instruments. They need to go in depth into the practice of subtractive synthesis, sampling, physical modeling, and FM synthesis.

With all these tools at your disposal, finding the right time and place to use them comes with experience. Studio tricks are used and overused. It is all about finding the right tool for the right job.

When applying these techniques, a successful engineer will listen back and evaluate the successfulness of the techniques used. When triggering sounds, listen to see if the drums sound natural within the mix. If they do not, do they still work within the track? Does the end result compete with commercial recordings? What will you do to improve the next project that you work on? There is always room for improvement; a successful engineer strives to improve the techniques that were both successful *and* unsuccessful. The true mark of a successful recording is the satisfaction of the client.

Is it possible to overproduce a recording? The answer to this is "of course." The end result needs to reflect the ultimate goals of the artist. If he or she is looking for a final product with consistent drums, perfect vocals, and everything rhythmically precise, then using all these production techniques is perfectly acceptable. If the client wants a natural-sounding recording, with fluctuations in tempo and pitch, then the use of the tools in this book should be more subtle. You certainly can add some pitch correction on the vocals, but it should not sound as severe as the pitch correction that is used on today's top-40 pop recordings.

Just like a professional recording studio, a home or project studio needs to be constantly upgraded and improved. Every year there are more and more tools available for both the professional and home studio, with most of these tools overlapping as the prices for various software instruments and plug-ins are relatively inexpensive, compared to their hardware counterparts.

The challenge for today's engineers is to be able to stay current with technology. There are many old-school engineers who still only record to analog tape and use production techniques that went along with it. There is nothing wrong with this, and you certainly cannot complain about these engineers who stay busy. Emerging engineers in today's marketplace need to gain a name for themselves to build a reputation and a steady stream of clientele.

Having great equipment does not necessarily make a great sound. It's the engineer who knows how to use the tools that makes the great-sounding records. A great engineer with cheap equipment can make a better-sounding record than a bad engineer with the best equipment.

Building up your skills as an engineer is a combination of becoming fluent with today's technology, learning how to work with clients and make them feel comfortable in the studio, and "building up your ears" as a professional. Learning technology does not take near as much time as it does to learn how to mix a great-sounding record. As with any profession, you will only get better with time. Just like musicians practice their instruments to get better, you need to practice working in the studio with the tools and techniques to become a better engineer. Also, never underestimate the amount that you can learn from watching another engineer work.

Index